T0146114

EARLY AMERICAN SCIENTIFIC AND TECHNICAL LITERATURE:

an annotated bibliography of books, pamphlets, and broadsides

by
MARGARET W. BATSCHELET

The Scarecrow Press, Inc.
Metuchen, N.J., & London
1990

British Library Cataloguing-in-Publication data available

Library of Congress Cataloging-in-Publication Data

Batschelet, Margaret.
    Early American scientific and technical literature : an annotated
bibliography of books, pamphlets, and broadsides / by Margaret W.
Batschelet.
        p.   cm.
    ISBN 0-8108-2318-7
    1. Science--Early works to 1800--Bibliography.   2. Engineering--
Early works to 1800--Bibliography.   3. United States--Imprints--
Early works to 1800--Bibliography.   I. Title.
Z7402.B38   1990
[Q157]
016.5--dc20   90-8095

To Bill, Josh, and Ben

Table of Contents

Introduction

It is usually accepted that the contributions of early American science were relatively modest compared with the work of European scientists. Raymond Stearns lists only fourteen early American scientists who "set forth scientific ideas illustrative of rationalizations and hypotheses based upon observations of data but transcending the data themselves" (679). Bernard Cohen finds only one scientist who contributed to what he terms "pure science": Benjamin Franklin (qtd. in Stearns 678). Yet even though their contributions were limited, the volume of material produced by early American scientific and technical writers is surprisingly large. Limiting my study to only books, pamphlets and broadsides, I have found over eight hundred works which fall into the broad category of early American scientific and technical literature.

This bibliography lists those scientific and technical books, pamphlets, and broadsides published in the United States between 1665 (the date of the earliest work, Samuel Danforth's *An Astronomical Description of the Late Comet or Blazing Star*) and 1799. The entries have been taken from a variety of sources including several general bibliographies and bibliographies covering limited scientific and technical subject matter. In the remainder of this Introduction, I will describe the divisions and contents of this bibliography, the material not included, the annotations for the entries, and other bibliographies consulted.

Divisions

I have divided the bibliography into three sections to simplify location of the entries. These sections are Medical Titles, Technical Science Titles, and Natural and Physical Science Titles.

Medical Titles include entries in medicine, pharmacy, or veterinary science.

Technical Science Titles include entries in agriculture, engineering, geography (including maps), manufacturing, mathematics, and statistics.

Natural and Physical Science Titles include entries in astronomy, botany, chemistry, geology, meteorology, physics, and zoology.

Obviously, there is some overlap among these sections. For example, Jefferson's *Notes on Virginia* is included in the *Technical Science* section because of its geographical content, but it might just as easily be included in the *Natural and Physical Science* section for its natural history material. Such attributions are often somewhat arbitrary, but I have attempted to place works in logical categories on the basis of their purpose and content.

Exclusions

Although I have attempted to include as much material as possible in this bibliography, it has been necessary to impose some limits on the content.

Because my primary interests here are books, pamphlets, and broadsides, I have excluded the large amount of scientific and technical material present in periodicals and almanacs. I hope to return to this material at a later time. I have also excluded most American reprints of foreign editions; the only exceptions to this exclusion are documents by American authors which were originally published by foreign presses (e.g., Zabdiel Boylston's *An Historical Account of the Small Pox Inoculation in New-England*, 1730).

I have included some government documents, both state and federal, but not all. I have included reports and monographs (e.g., Jefferson's *Report of the Secretary of State on the Subject of Establishing a Uniformity in the Weights, Measures and Coins of the United States*, 1790), and collections of materials (e.g. *Several Methods of Making Salt-Petre* published by the Continental Congress in 1775). However, I have excluded all of the many legislative acts pertaining to scientific and technical topics.

Annotations

At the end of each entry I have included a brief annotation, chiefly to indicate the content of the work. I have also attempted to indicate the importance of some prominent works and in doing so I have used several sources. The abbreviations for my most common references are

BDAS - *The Biographical Dictionary of American Science*
DAB - *The Dictionary of American Biography*
DAMB - *The Dictionary of American Medical Biography*
DSB - *The Dictionary of Scientific Biography*

A list of other reference works referred to in the annotations can be found at the end of this introduction.

Other Bibliographies

Following each entry is a list of other bibliographies in which the work is cited, along with the reference number used within the bibliography (if one is given) or the page number (if no reference number is listed).

Most of the entries in this bibliography can also be found in either Charles Evans' *American Bibliography* or Robert Bristol's *Supplement to Charles Evans' American Bibliography*. In turn, most entries from Evans and Bristol are included in the Readex Early American Imprints Series 1 (available on microfiche or on microprint in an earlier version) and are thus easily accessible to many American readers. Because of the wide availability of the Readex series and its value as a research tool, I have indicated which entries in this bibliography are *not* included in Readex and, if possible, why they have been excluded (e.g., the only recorded copy is missing).

Unless otherwise indicated, all entries with entry numbers from Evans can be found in Readex under those entry numbers. In the case of the entries taken from Robert Bristol, I have used Bristol's system of a microprint number ("mp") for the Readex entry as well as the entry number under which the work can be found in Bristol.

A complete list of all the bibliographies cited is given at the end of this introduction.

Acknowledgements

In compiling this bibliography I have made use of several library collections, but I am particularly indebted to the John Peace Library at the University of Texas at San Antonio and the Elizabeth Coates Maddux Library at Trinity University. Both libraries provided valuable information and the staffs of both institutions were unstintingly helpful.

The text of this bibliography was printed using a Macintosh computer and Apple Laser Writer, and for the use of that equipment I must express my gratitude to the Division of English, Classics, and Philosophy at the University of Texas at San Antonio. The generosity of the division in making available its desktop publishing system has simplified my task immeasurably.

Finally, I wish to thank both my husband, Bill, and my sons, Josh and Ben, for their patience while I was engrossed in early American science. Without their support and understanding, this bibliography could not have been completed.

Margaret Batschelet
University of Texas at San Antonio
San Antonio, Texas

References

Binger, Carl. *Revolutionary Doctor: Benjamin Rush, 1746-1813.* New York: W.W. Norton and Co., Inc., 1966.

D'Elia, Donald J. "Benjamin Rush: Philosopher of the American Revolution." *Transactions of the American Philosophical Society* 64 (1974): 5-107.

The Dictionary of American Biography. Eds. Allen Johnson and Dumas Malone. New York: Charles Scribner's Sons, 1937. (DAB)

The Dictionary of Scientific Biography. Ed. Charles Coulston Gillispie. New York: Charles Scribner's Sons, 1972. (DSB)

Eliott, Clark A. *Biographical Dictionary of American Science: The Seventeenth through the Nineteenth Centuries.* Westport, CT: Greenwood Press, 1979. (BDAS)

Greene, John C. *American Science in the Age of Jefferson.* Ames, IA: Iowa State University Press, 1984.

Hawke, David Freeman. *Benjamin Rush: Revolutionary Gadfly.* New York: Bobbs Merrill Co., Inc., 1971.

Hindle, Brooke. *The Pursuit of Science in Revolutionary America: 1735-1789.* Chapel Hill: University of North Carolina Press, 1956.

A History of Mathematical Education in the United States and Canada. 32nd Yearbook. Washington, DC: National Council of Teachers of Mathematics, 1970.

Tregle, Joseph G., Jr., ed. *An Historical Narrative and Topographical Description of Louisiana and West Florida.* By Thomas Hutchins. Facsimile reproduction of the 1784 edition. Gainesville: University of Florida Press, 1968.

Judson, Louis V. *Weights and Measures Standards of the United States: A Brief History.* National Bureau of Standards Special Publication 447. 1963. Washington, DC: GPO, 1976.

Karpinski, Louis C. "Colonial American Arithmetics." *Bibliographical Essays: A Tribute to Wilberforce Eames.* Essay Index Reprint Series. 1924. Freeport, NY: Books for Libraries Press, Inc., 1967.

Lemay, J.A. Leo. *Ebenezer Kinnersley: Franklin's Friend.* Philadelphia: University of Pennsylvania Press, 1964.

Niltz, John A. *The Evolution of American Secondary School Textbooks.* Rutland, VT: Charles E. Tuttle Co., 1966.

Silverman, Kenneth. *The Life and Times of Cotton Mather.* New York: Harper and Row, 1984.

Stearns, Raymond P. *Science in the British Colonies of America.* Urbana: University of Illinois Press, 1970.

Viets, Henry R.,ed. *A Brief Rule to Guide the Common People of New England How to Order Themselves and Theirs in the Small Pocks or Measels.* By Thomas Thacher. Bibliotheca Medica Americana, ser. 4, 1. Publication of the Institute of the History of Medicine. Baltimore: Johns Hopkins Press, 1937.

Wilson, M.L. "Survey of Scientific Agriculture." *The Early History of Science and Learning in America: Proceedings of the American Philosophical Society* 86 (1942): 52-62.

Winslow, Charles-Edward Amory. *The Conquest of Epidemic Disease: A Chapter in the History of Ideas.* 1943. New York: Hafner Publishing Co., 1967.

Bibliographies Cited

Austin, Robert B. *Early American Medical Imprints: A Guide to Works Printed in the United States, 1668-1820*. Washington, DC: United States Department of Health, Education, and Welfare, Public Health Service, 1961.

Bristol, Robert P. *Supplement to Charles Evans' American Bibliography*. Charlottesville: University Press of Virginia, 1970.

Evans, Charles. *American Bibliography*. Volumes 1-12. New York: Peter Smith, 1941.

Guerra, Francisco. *American Medical Bibliography, 1639-1783*. New York: Harper, 1962.

Hazen, Robert M. and Margaret Hindle Hazen. *American Geological Literature, 1669 to 1850*. Stroudsburg, PA: Dowden, Hutchinson and Ross, Inc., 1980.

Karpinsky, Louis C. *Bibliography of Mathematical Works Printed in America through 1850*. Ann Arbor: University of Michigan Press, 1940.

Meisel, Max. *A Bibliography of American Natural History: The Pioneer Century, 1769-1865*. 3 Vols. 1929. New York: Hafner Publishing Co., 1967.

Miller, C. William. *Benjamin Franklin's Philadelphia Printing, 1728-1766: A Descriptive Bibliography*. Philadelphia: American Philosophical Society, 1974.

Shipton, Clifford K. *The American Bibliography of Charles Evans, Volume 13, 1799-1800*. 1955. Worcester, MA: American Antiquarian Society, 1962 (cited as Evans).

Wheat, James Clement and Christian F. Brun. *Maps and Charts Published in America before 1800: A Bibliography*. New Haven: Yale University Press, 1969.

Medical Titles

1677

1 **Thacher, Thomas (1620-1678).** *A Brief Rule to Guide the Common People of New England How to Order Themselves and Theirs in the Small Pocks [sic] or Measels [sic].* Boston: John Foster, 1677. Evans: 242 Austin: 1881 Guerra: a-3. First treatment of smallpox written in America. Thacher was minister of Third Church in Boston; Viets calls him "the best example of a preacher-physician of his time." His work is based entirely on Thomas Sydenham's *Methodus Curandi Febres* (1666), sometimes word for word. It includes a definition of the disease, a description of its symptoms and treatment.

1721

2 **Boylston, Zabdiel (1680-1766) [and Cotton Mather (1662-1728)].** *Some Account of What Is Said of Inoculating or Transplanting the Small Pox. By the Learned Dr. Emanuel Timonius, and Jacobus Pylarinus with Some Remarks Thereon. To Which Are Added, a Few Queries in Answer to the Scruples of Many about the Lawfulness of This Method.* Boston: S. Gerrish, 1721. Evans: 2206 Austin: 1231 Guerra: a-51. Joint project of Boylston and Mather prepared as a defense against the anti-inoculation forces. Parts I and II are credited to Mather; Part III is credited to Boylston.

3 **Colman, Benjamin (1673-1747).** *Some Observations on the New Method of Receiving the Small Pox by Ingrafting or Inoculating. Containing Also the Reasons, Which First Induc'd Him to, and Have Since Confirm'd Him in, His Favourable Opinion of It.* Boston: B. Green, 1721. Evans: 2211 Austin: 504 Guerra: a-52. Another defense of inoculation. Colman was minister of Brattle Street Church as well as a Fellow and Overseer of Harvard. According to Guerra this work caused "repercussions" when it was published in England in 1722; it "clarifies many points concerning the early introduction and practice of inoculation in New England."

4 **[Cooper, William (1694-1743)].** *A Letter to a Friend in the Country, Attempting a Solution of the Scruples and Objections of a Conscientious or Religious Nature, Commonly Made against the New Way of Receiving the Small Pox.* Boston: S. Kneeland, 1721. Evans: 2247 Austin: 538 Guerra: a-53. Defense of inoculation on religious grounds. The tract was originally published anonymously, republished in London in 1722, and published in a third edition in Boston in 1730. In this edition Cooper signed a preface to the reader.

5 **[Douglass, William (1691-1752)].** *A Letter from One in the Country, to His Friend in the City: In Relation to Their Distresses Occasioned by the Doubtful and Prevailing Practice of the Inocculation [sic] of the Small Pox.* Boston: Nicholas Boone, 1721. Evans: 2229 Austin: 1142 Guerra: a-54. Anti-inoculation tract originally published

anonymously. Douglass was the leader of the medical attack on inoculation; his arguments are partly procedural, but he also includes religious objections.

6 **[Grainger, Henry].** *The Imposition of Inoculation As a Duty Religiously Considered in a Letter to a Gentleman in the Country Inclin'd to Admit It.* Boston: Nicholas Boone and John Edwards, 1721. Evans: 2222 Austin: 831 Guerra: a-55. Anti-inoculation tract. Grainger attacks inoculation on religious grounds, as well as on the grounds that evidence in favor of it has been gained from "Africans" and "Mahometans."

7 **Mather, Increase (1639-1723).** *Several Reasons Proving the Inoculating or Transplanting the Small Pox Is a Lawful Practice, and That It Has Been Blessed by God for the Saving of Many a Life.* Boston: S. Kneeland, 1721. Evans: 2258 Austin: 1233 Guerra: a-64. Increase Mather's defense of inoculation; it includes a scathing attack on opponents of inoculation. According to Guerra "Sentiments on the Small Pox Inoculated" by Cotton Mather is printed on the verso.

8 **Mather, Increase (1639-1723).** *Some Further Account from London of the Small Pox Inoculated, with Some Remarks on a Late Scandalous Pamphlet Entitled, Inoculation of the Small Pox As Practis'd in Boston, &c.* Boston: J. Edwards, 1721. Evans: 2259 Austin: 1234 Guerra: a-65. Mainly taken up with an abstract of a lecture by Dr. Walter Harris, "De Inoculatione Variolarum," which he delivered to the Royal College of Physicians in London. It also includes both a summary of a communication from Timonius to Woodward describing inoculation in Constantinople and an attack on Douglass.

9 **Williams, John.** *Several Arguments, Proving That Inoculating the Small Pox Is Not Contained in the Law of Physick, Either Natural or Divine, and Therefore Unlawful, Together with a Reply to Two Short Pieces, One by the Rev. Dr. Increase Mather, and Another by an Anonymous Author, Intituled [sic], Sentiments on the Small Pox Inoculated, and Also a Short Answer to a Late Letter in the New England Courant.* Boston: J. Franklin, 1721. Evans: 2307 Austin: 2058 Guerra: a-66. Another attack on inoculation on religious and legal grounds. According to Silverman, Williams was the owner of a "tobacco cellar" in Boston, "where he dispensed tobacco, drugs, and free medical advice." He was a leader of the anti-clerical group in the inoculation controversy.

1722
10 **[Douglass, William (1691-1752)].** *The Abuses and Scandals of Some Late Pamphlets in Favour of Inoculation of the Small Pox, Modestly Obviated, and Inoculation Further Consider'd in a Letter to A[lexander] S[tewart] M.D. & F.R.S. In London.* Boston: J. Franklin, 1722. Evans: 2331 Austin: 685 Guerra: a-70. Douglass' answer to

Increase and Cotton Mather, chiefly an attack on the latter. It was published anonymously; Douglass identified himself in the London edition. The original recipient was Alexander Stewart.

11 **[Douglass, William (1691-1752)].** *Inoculation of the Small Pox As Practised in Boston, Consider'd in a Letter to A[lexander] S[tewart] M.D. & F.R.S. In London.* Boston: J. Franklin, 1722. Evans: 2332 Austin: 687 Guerra: 69. Douglass' most complete statement of his opposition to inoculation. He attacks many pro-inoculation figures, including Boylston, C. Mather, and Colman. Guerra calls it "a clear and unvarnished historical account of inoculation in New England, a statement of the arguments used by the inoculators, and finally some remarks on the practice."

12 **Douglass, William (1691-1752).** *Postscript to Abuses &c. Obviated, Being a Short and Modest Answer to Matters of Fact Maliciously Misrepresented in a Late Doggerel Dialogue.* Boston: J. Franklin, 1722. Evans: 2333 Austin: 688 Guerra: a-71. An answer to Greenwood's burlesque (see 14). Douglass accuses Greenwood both of inaccuracy and poor literary style.

13 *A Friendly Debate; or, a Dialogue between Rusticus and Academicus about the Late Performance of Academicus.* Boston: J. Franklin, 1722. Evans: 2386 Austin: 1696 Guerra: a-73. A rejoinder to Greenwood (see 14), but it also includes an attack on John Williams and is dedicated to Cotton Mather. According to Austin it was sometimes attributed to Samuel Mather, who denied authorship.

14 **[Greenwood, Isaac (1702-1745)].** *A Friendly Debate; or, a Dialogue between Academicus and Sawny and Mundungus, Two Eminent Physicians, about Some of Their Late Performances.* Boston: 1722. Evans: 2339 Austin: 840 Guerra: a-72. Burlesque dialogue between Academicus (Boylston and Mather), Sawny (Douglass), and Mundungus (Williams). Greenwood ridicules Douglass on the basis of his reasoning, his skill as a physician, and his broad Scots dialect.

15 **[Mather, Cotton (1662-1728)].** *The Angel of Bethesda, Visiting the Invalids of a Miserable World.* New London: Timothy Green, 1722. Evans: 2352 Austin: 1228 Guerra: a-74. Chapter five of Mather's unpublished medical treatise by the same title. It discusses the influence of what Mather calls "Nishmath-Chijim or Breath of Life" on health. Guerra describes it as an early example of psychosomatic medicine.

16 **[Mather, Cotton (1662-1728) and others].** *A Vindication of the Ministers of Boston from the Abuses and Scandals, Lately Cast upon Them in Diverse Printed Papers.* Boston: B. Green, 1722. Evans: 2396 Austin: 1980 Guerra: a-76. Defense of inoculation and attack on the anti-clerical party. The ministers cited are Cotton and

Increase Mather, Benjamin Colman, and William Cooper among others. Not all of the authors are known, but Mather contributed to the work.

17 **Williams, John (1664-1729).** *An Answer to a Late Pamphlet, Intitled [sic]: A Letter to a Friend in the Country, Attempting a Solution of the Scruples and Objections of a Conscientious or Religious Nature Commonly Made against the New Way of Receiving the Small Pox, by a Minister of Boston, Together with a Short History of the Late Divisions among Us in Affairs of State, and Some Account of the First Cause of Them.* Boston: J. Franklin, 1722. Evans: 2407 Austin: 2057 Guerra: a-78. Another anti-clerical attack. Only the first seven pages are taken up with medical arguments; the remainder concerns economic and civil questions.

1730

18 **Boylston, Zabdiel (1680-1766).** *An Historical Account of the Small-Pox Inoculation in New England upon All Sorts of Persons, Whites, Blacks, and of All Ages and Constitutions, with Some Account of the Nature of the Infection in the Natural and Inoculated Way, and Their Different Effects on Human Bodies, with Some Short Directions to the Less Experienced in This Method of Practice.* London. Reprinted Boston: S. Gerrish and T. Hancock, 1730. Evans: 3259 Austin: 263 Guerra: a-110. Boylston's first publication of his results under his own name; the DAB terms it "in every way a masterly clinical presentation--the first of its kind from an American physician." Boylston conducted the first large-scale inoculation (247 people) in the West; according to Richard Shryock (DAMB) it was "the chief American contribution to medicine before the mid-nineteenth century."

19 **Douglass, William (1691-1752).** *A Dissertation Concerning Inoculation of the Small Pox, Giving Some Account of the Rise, Progress, Success, Advantages and Disadvantages of Receiving the Small Pox by Incisions, Illustrated by Sundry Cases of the Inoculated.* Boston: D. Henchman and T. Hancock, 1730. Evans: 3274 Austin: 686 Guerra: a-115. Douglass' acceptance of inoculation; the work is based on English results rather than those of Boylston in New England, and it contains an attack on Boylston. According to Guerra it "constitute[s] the first comparative study of overall mortality and that from smallpox in America."

20 **Douglass, William (1691-1752).** *A Practical Essay Concerning the Small Pox.* Boston: D. Henchman and T. Hancock, 1730. Evans: 3275 Austin: 689 Guerra: a-117. A clinical description of the smallpox epidemic of 1721. Douglass uses contemporary medical terminology and the publisher attaches a glossary because Douglass refuses to "debase" his treatise by "using of a vulgar stile with low Circumlocutions."

21 *A Letter to Doctor Zabdiel Boylston Occasion'd by a Late Dissertation Concerning Inoculation.* Boston: D. Henchman and T. Hancock, 1730. Evans: 3296 Austin: 1143 Guerra: a-116. A repudiation of Douglass *Dissertation* (see 19) and a defense of Boylston and Mather.

1732

22 Harward, Thomas (1700-1736). *Electuarium Novum Alexipharmacum: Or a New Cordial, Alexiterial, and Restorative Electuary, Which May Serve for a Succedaneum to the Grand Theriaca Andromachi.* Boston: B. Green, 1732. Evans: 3549 Austin: 889 Guerra: a-127. Description of various medicinal herbs and herbal mixtures used in the preparation of Harward's electuary (a mixture of a powdered drug in honey or syrup). The theriaca was a medieval medicine supposedly for the treatment of poisonous animal bites. Guerra states that the work "reflects the influence of the European pharmacopoeias." Harward was a lecturer at King's Chapel in Boston and Licenciate of the Royal College of Physicians.

23 Walton, John. *An Essay on Fevers, the Rattles, and Canker.* Boston: T. Fleet, 1732. Evans: 3614 Austin: 1996 Guerra: a-130. Description of the causes of fever, its symptoms, duration and treatment. Walton is highly critical of American physicians; he was admitted to the Royal College of Physicians in 1750.

1734

24 [Tennent, John (1700-c.1760)]. *Every Man His Own Doctor or the Poor Planter's Physician, Plain and Easy Means for Persons to Cure Themselves of All or Most of the Distempers Incident to This Climate, and with Very Little Charge, the Medicines Being Chiefly of the Growth and Production of This Country.* Williamsburg and Annapolis: 1734. Evans: 3843 Austin: 1870 Guerra: a-139. Famous popular medical guide; it went through six editions (one in German). Benjamin Franklin reprinted Tennent in 1734 with a warning against Tennent's prescription of Ipecac. Authorship is usually attributed to Tennent, but according to Austin the attribution is uncertain.

1736

25 Douglass, William (1697-1752). *The Practical History of a New Epidemical Eruptive Miliary Fever, with an Angina Ulcusculosa Which Prevailed in Boston New England in the Years 1735 and 1736.* Boston: Thomas Fleet, 1736. Evans: 4012 Austin: 690 Guerra: a-145. Description of an epidemic of scarlet fever, including symptoms and treatment. The DAMB describes it as an "outstanding study of the Boston scarlet-fever epidemic of 1735-36 and the first adequate description of that disease in English."

26 [Fitch, Jabez (1672-1746)]. *An Account of the Numbers That Have Died of the Distemper in the Throat within the Province of New-Hampshire, with Some Reflections Thereon.* Boston: Eleazer Russel,

1736. Evans: 4014 Austin: 775 Guerra: a-146. Statistical account of the mortality rates during a New Hampshire diphtheria epidemic, with a theological interpretation. Fitch was a minister in Portsmouth.

27 **Tennent, John (1700-c.1760)].** *An Essay on Pleurisy.* Williamsburg: William Parks, 1736. Evans: 4085 Austin: 1868 Guerra: a-151. Description of the disease, with probable causes and suggestions for cure, chiefly the use of "Seneca rattlesnake root." According to the DAB, Tennent's endorsement of rattlesnake root brought him "widespread notoriety."

1737

28 **Henchman, Daniel (1689-1791).** *Reverend Sir, Boston, March 12, 1736,7.* [Boston: 1737]. mp: 40116 Austin: 901 Bristol:1002. Circular letter sent to New Hampshire ministers requesting statistics on deaths from diphtheria from 1735-36; Henchman proposed to publish the statistics as a supplement to Fitch (see 26).

1738

29 **Brown, John (1696-1752).** *The Number of Deaths in Haverhill and Also Some Comfortable Instances Thereof among the Children under the Late Distemper of the Throat, with an Address to the Bereaved.* Boston: S. Kneeland and T. Green, 1738. Evans: 4128 Guerra: a-156. Statistical table listing families and deaths from diphtheria per day, accompanied by a sermon.

30 *Letter Concerning the Proper Treatment of the Smallpox by Laicus.* Charleston, SC: Lewis Timothy, 1738. mp: 40145 Not included in Readex Bristol:1028. Apparently an inoculation essay; no copy has been found.

1739

31 **Killpatrick, James (d. 1770).** *A Full and Clear Reply to Doctor Thomas Dale, Wherein the Real Impropriety of Blistering with Cantharides in the First Fever of the Small Pox is Plainly Demonstrated, with Some Diverting Remarks on the Doctor's Great Consistence, and Conquisite Attainments in Physick and Philology.* Charlestown: 1739. Evans: 4373 Austin: 1096 Guerra: a-175. Third in a series of pamphlets in a dispute between Killpatrick and Dale, and the only one of the series with existing copies. The dispute centers on the treatment of a case of smallpox. The pamphlet closes with statistics showing lower mortality among the inoculated than the non-inoculated. Killpatrick later published two studies of inoculation in London.

32 **[Mather, Cotton (1662-1728)].** *A Letter about a Good Management under the Distemper of the Measles, at This Time Spreading in the Country, Here Published for the Benefit of the Poor, and Such As May Want the Help of Able Physicians.* Boston: 1739. Evans: 4376 Austin: 1230 Guerra: a-176. Description of treatment,

including herbal medicines and home remedies; intended for families of patients. Published posthumously.

1740

33 **Dickinson, Jonathan (1688-1747).** *Observations on That Terrible Disease Vulgarly Called the Throat Distemper, with Advice As to the Method of Cure.* Boston: S. Kneeland and T. Green, 1740. Austin: 670 Guerra: a-177. Description of "eruptive miliary fever" (probably diphtheria), with symptoms and cure. Dickinson was the first president of Princeton, a minister in Elizabethtown, New Hampshire, and a physician. Guerra terms it "a careful description of symptoms."

1745

34 **Berkeley, George [Colden, Cadwallader (1688-1766)].** *A n Abstract from Dr. Berkely's [sic] Treatise on Tar-Water, with Some Reflexions Thereon, Adapted to Diseases Frequent in America.* New York: J. Parker, 1745. Evans: 5539 Austin: 492 Guerra: a-211. Adaptation of Berkeley for lay readers; the first nine pages are an abstract, followed by applications of tar-water, along with herbal remedies, to American diseases. Attributed to Colden by S. Jarcho.

35 **Cadwalader, Thomas (1707-1779).** *An Essay on the West-India Dry Gripes; with the Method of Preventing and Curing That Cruel Distemper, to Which Is Added, an Extraordinary Case in Physick.* Philadelphia: B. Franklin, 1745. Evans: 5553 Austin: 380 Miller: 369 Guerra: a-213. Description of "dry gripes," form of lead poisoning caused by lead pipes used in distilling rum. The essay also includes a description of a case of osteomalacia based on an autopsy performed by Cadwalader in 1742, one of the earliest recorded autopsies in America.

1750

36 **Thomson, Adam (d. 1767).** *A Discourse on the Preparation of the Body for the Small Pox and the Manner of Receiving the Infection.* Philadelphia: Franklin and Hall, 1750. Evans: 6617 Austin: 1899 Miller: 518 Guerra: a-240. Originally delivered as an address to the trustees of the Philadelphia Academy. The essay was controversial because Thomson advocated the use of mercury and antimony as preparations for inoculation and because it contained criticism of American medicine.

1751

37 **Hamilton, Alexander (1712-1756).** *A Defense of Dr. Thomson's Discourse on the Preparation of the Body for the Small Pox, and the Manner of Receiving the Infection.* Philadelphia: W. Bradford, 1751. Evans: 6689 Austin: 859 Guerra: a-243. Answer to John Kearsley's attack on Thomson (see 38). Hamilton is better known as a diarist than as a physician; this pamphlet is his only medical publication.

38 Kearsley, John (1684-1772). *A Letter to a Friend Containing Remarks on a Discourse Proposing a Preparation of the Body for the Small Pox and the Manner of Receiving the Infection, with Some Practical Hints Relating to the Cure of the Dumb Ague, Long Fever, the Bilious Fever, and Some Other Fevers, Incidental to This Province.* Philadelphia: Franklin and Hall, 1751. Evans: 6697 Austin: 1093 Miller: 532 Guerra: a-244. Answer to Thomson (see 36). Kearsley does not attack inoculation itself, merely Thomson's methods. The essay includes a clinical discussion of the case in question; Guerra describes it as "the first example of medical expert testimony in the colonies."

39 Short, Thomas (1690-1772) and Bartram, John (1699-1777). *Medicina Britanica or a Treatise on Such Physical Plants As Are Generally to Be Found in the Fields or Gardens in Great Britain: Containing a Particular Account of Their Nature, Virtues, and Uses. With a Preface by Mr. John Bartram, Botanist of Pennsylvania, and His Notes throughout the Work, Shewing the Places Where Many of the Described Plants Are to Be Found in These Parts of America, Their Differences in Name, Appearance, and Virtues from Those of the Same Kind in Europe; and an Appendix, Containing a Description of a Number of Plants Peculiar to America, Their Uses, Virtues, &c.* London. Reprinted Philadelphia: Franklin and Hall, 1751. Evans: 6783 Austin: 1743 Guerra: a-246 Meisel: 3:345 Miller: 546. Popular British pharmacopoeia. Bartram's additions give the locations of individual plants in America and different names, if any, as well as occasional identifying characteristics, such as smell. The appendix gives "Descriptions, Virtues, and Uses" of North American plants, as well as directions for preparation of herbal medicines.

1752

40 Williams, Nathaniel (1675-1738). *The Method of Practice in the Small Pox with Observations on the Way of Inoculation.* Boston: Kneeland, 1752. Evans: 6947 Austin: 2062 Guerra: a-262. Description of treatment for smallpox. It includes directions for inoculation.

1754

41 [Briant, Timothy]. *A Receipt from Middleborough in the Province of Massachusetts Concerning the Canker or Throat Distemper.* Boston: D. Fowle, [1754]. mp: 40682 Bristol:1643 Austin: 273 Guerra: a-268. Directions for preparations of home remedies for "throat distemper" (probably diphtheria).

42 [Franklin, Benjamin (1706-1790)]. *Some Account of the Pennsylvania Hospital from Its First Rise to the Beginning of the Fifth Month, Called May, 1754.* Philadelphia: Franklin and Hall, 1754. Evans: 7197 Austin: 794. Miller: 587 Guerra: a-272. History of the Pennsylvania Hospital from 1750 onward. It includes petitions,

addresses and other related documents, as well as rules adopted, patient
statistics, budget, and lists of physicians and surgeons. The
Pennsylvania Hospital was the first in the country "exclusively for the
care of the sick" (DAMB).

1756

43 **Macleane, Lauchlin (c.1728-1777).** *An Essay on the
Expediency of Inoculation and the Seasons Most Proper for It.*
Philadelphia: William Bradford, 1756. Evans: 7701 Austin: 1177
Guerra: a-283. Discussion of contemporary theories concerning
smallpox and inoculation. It includes tables of mortality figures.
Macleane declares his agreement with Thomson (see 36).

1759

44 **Redman, John (1722-1808).** *A Defence of Inoculation.*
Philadelphia: 1759 Evans: 8477 Not included in Readex Austin: 1588
Guerra: a-301. Pro-inoculation pamphlet, apparently originally printed
as a letter in the *Pennsylvania Gazette.* No copy is known to exist.
Redman was a prominent physician, the teacher of Benjamin Rush and
John Morgan, among others.

1760

45 *Directions Concerning Inoculation, Chiefly Collected from the Late
Pieces on That Subject with Instructions How to Prepare Those Who
Are Soonest Likely to Take the Small-Pox in the Natural Way.*
Philadelphia: Franklin and Hall; Boston: Mecom, 1760. Evans: 8581
Austin: 673 Guerra: a-304. Directions for inoculators, including
medication. The pamphlet was intended for both physicians and
laymen. Guerra suggests it may actually be the Redman pamphlet (see
44).

46 **[Kearsley, John (1684-1772)].** *The Case of Mr. T[homas]
L[awrence] with Regard to the Method Pursued Therein by J[ohn]
K[earsley] Surgeon, with the Uncommon Treatment the Said J[ohn]
K[earsley] Hath Met with in His Proceedure [sic] Therein.*
Philadelphia: 1760. Evans: 8630 Austin: 1760. Guerra: a-313.
Kearsley's defense of his treatment of Thomas Lawrence, who
developed an infection after Kearsley treated him for a tumor (Guerra:
"fistula lachrymalis"). Lawrence had apparently challenged Kearsley's
competence.

1761

47 **[Gardiner, Silvester (1708-1786)].** *To the Freeholders and
Other Inhabitants of the Town of Boston, in Town Meeting Assembled,
March, 1761.* [Boston, 1761]. Evans: 8862 Austin: 813 Guerra: a-
321. Proposal for building an inoculation hospital in Boston.
Apparently it was unsuccessful.

48 Pennsylvania Hospital. *Continuation of the Account of the Pennsylvania Hospital, from the First of May 1754, to the Fifth of May, 1761.* Philadelphia: Franklin and Hall, 1761. Evans: 8972 Austin: 794 Miller: 764 Guerra: a-324. Continuation of the chronicle of the hospital, including new construction, capital, expenses, diseases treated, and managers. These reports continued periodically through the nineteenth century. This installment was not written by Franklin (see 42) but by a "Committee on Publication" chaired by Samuel Rhoads.

1763
49 Pennsylvania Hospital. *Rules for Admission and Discharge of Patients.* [Philadelphia: B. Franklin and D. Hall, 1763]. Bristol: 1587 Miller: 797. Hospital rules; also included in *Some Account of the Pennsylvania Hospital* (see 42) .

1764
50 Garden, Alexander (1730-1791). *An Account of the Medical Properties of the Virginia Pink-Root.* Charlestown, S.C.: Peter Timothy, 1764. Evans: 9675 Not in Readex Guerra: a-346. Description of "vermifugal properties" of pinkroot (DAB). Garden, an important early botanist, promoted the medicinal use of pinkroot.

1765
51 Morgan, John (1735-1789). *Apology for Attempting to Introduce the Regular Practice of Physic in Philadelphia.* Philadelphia: William Bradford, 1765. mp: 41569 Bristol: 2601 Austin: 1333 Guerra: a-368. Proposals for reform of medical education and practice, including the institution of a governing board and regulations for controlling practitioners. Morgan's proposals caused some controversy among Philadelphia physicians.

52 Morgan, John (1735-1789). *A Discourse upon the Institution of Medical Schools in America; Delivered at a Public Anniversary Commencement, Held in the College of Philadelphia, May 30 and 31, 1765.* Philadelphia: William Bradford, 1765. Evans: 10082 Austin: 1335 Guerra: a-367. Proposals for founding a medical school in Philadelphia, with a systematic description of a course of studies. It includes a general survey of medicine in America. This proposal is frequently bound with the *Apology* (see 51). The DAB calls it "a classic." Morgan was one of the founders of the University of Pennsylvania medical school.

1767
53 Chalmers, Lionel (1715-1777). *An Essay on Fevers, More Particularly Those of the Common Continued and Inflammatory Sorts, Wherein a New and Successful Method Is Proposed for Removing Them Speedily.* Charleston: Robert Wells, 1767. Evans: 10575 Austin: 435 Guerra: a-392. Discussion of the cause and treatment of fevers, reporting statistical studies by both Chalmers and John Lining.

A second edition appeared in London in 1768, along with a later German translation.

1769

54 Bard, Samuel (1742-1821). *A Discourse upon the Duties of a Physician, with Some Sentiments on the Usefulness and Necessity of a Public Hospital, Delivered before the President and Governors of King's College, at the Commencement, Held on the 16th of May, 1769.* New York: J. Robertson, 1769. Evans: 11168 Austin: 124 Guerra: a-423. Address delivered at the first commencement of the first medical school in New York (at King's College, later part of Columbia University). Bard was professor of the theory and practise of physic and a prime mover in organizing the medical school (the second in the colonies after the University of Pennsylvania).

55 Hamilton, Thomas (fl. 1769). *Some Account of the Small Pox in the Town of Chatham, in the Year 1766.* n.p.: 1769. mp: 41941 Bristol: 3008 Guerra: a-431. Description of a smallpox epidemic; the pamphlet has a largely religious orientation. Guerra cites the "good description of the case of Mr. Stephen Ryder."

56 [Kearsley, John (1684-1772)]. *Observations on the Angina Maligna, or the Putrid and Ulcerous Sore Throat, with a Method of Treating It.* Philadelphia: William and Thomas Bradford, 1769. Evans: 11449 Austin: 1094 Guerra: a-434. Discussion of diphtheria; it includes a condemnation of bleeding as a treatment. According to Guerra it "represents the treatment which was accepted until comparatively recently."

57 Middleton, Peter (d.1781) *A Medical Discourse or an Historical Inquiry into the Ancient and Present State of Medicine, the Substance of Which Was Delivered at Opening the Medical School, in the City of New York.* New York: Hugh Gaine, 1769. Evans: 11338 Austin: 1298 Guerra: a-436. Address delivered at the opening of the King's College medical school; it includes a short history of medicine and of medical education. It displays "considerable familiarity with medical history, but gives little on the American situation" (DAB).

1770

58 Dabney, Nathaniel. *Dr. Stoughton's Elixir Magnum Stomachicum.* [Salem: c.1770]. mp: 42032 Not included in Readex Bristol: 3108. Apparently an advertisement.

59 [Sparhawk, John]. *Essentia Euphragiae.* [Philadelphia?: c.1770] mp: 42044 Bristol: 3110. Description of eye medicine prepared and sold by Sparhawk.

1771

60 Bard, Samuel (1742-1821). *An Enquiry into the Nature, Cause, and Cure of the Angina Suffocativa, Sore Throat Distemper, As It Is Commonly Called by the Inhabitants of this City and Colony.* New York: S. Inslee and A. Car, 1771. Evans: 11977 Austin: 125 Guerra: a-465. Description of diphtheria, chiefly in children. Guerra calls the pamphlet "one of the best descriptions written of the symptoms of the diphtheria which had caused so many deaths over this past century." Abraham Jacobi terms it "correct and beautiful" (DAB).

61 Elmer, Jonathan (1745-1817). *Dissertatio Medica Inauguralis de Sitis in Febribus Causis et Remedis.* Philadelphia: Henry Miller, 1771. Evans: 12035 Austin: 724 Guerra: a-472. One of the first medical theses at the University of Pennsylvania medical school; on fevers. In Latin. Elmer went on to become a prominent doctor and member of the American Philosophical Society.

62 Kissam, Samuel (b. 1745). *An Inaugural Essay on the Anthelmintic Quality of the Paseolus Zuratensis Siliqua Hirsuta, or Cow-Itch.* New York: S. Inslee and A. Car, 1771. Evans: 12091 Austin: 1099 Guerra: a-474. First medical thesis in New York; for degree at King's College (Columbia). "Cow itch" is a tropical vine which, mixed with honey, served as a vermifuge. Kissam describes the symptoms of worm infestation in children and his results in using cow itch as a treatment.

63 New York (City) Hospital. *Charter for Establishing an Hospital in the City of New York.* New York: H. Gaine, 1771. Evans: 12161 Austin: 1795 Guerra: a-477. Description of officers of the New York Hospital with rules governing elections and procedures.

64 Potts, Jonathan (1745-1781). *Dissertatio Medica Inauguralis de Febribus Intermittentibus, Potentissimum Tertianis.* Philadelphia: John Dunlap, 1771. Evans: 12204 Austin: 1561 Guerra: a-481. Medical thesis for the University of Pennsylvania; a study of malaria. In Latin.

65 Tilton, James (1745-1822). *Dissertatio Medica Inauguralis de Hydrope.* Philadelphia: William and Thomas Bradford, 1771. Evans: 12242 Austin:1908 Guerra: a-483. Another early dissertation for the University of Pennsylvania medical school; study of dropsy. In Latin. Tilton was later Surgeon General.

66 Way, Nicholas (c.1750-1797). *Dissertatio Medica Inauguralis de Variolarum Insitione.* Philadelphia: Henry Miller, 1771. Evans: 12275 Austin: 2019 Guerra: a-486. Dissertation for the University of Pennsylvania medical school; on inoculation. In Latin.

1772

67 Bass, Robert. *Dr. Keyser's Famous Pills, Imported and Warranted Genuine by Robert Bass.* [Philadelphia: William and Thomas Bradford, 1772]. Evans: 42313 Bristol: 3424 Guerra: a-509. Broadside advertisement for pills (according to Guerra, a mercury preparation used in treating venereal disease, rheumatic complaints, and yaws), with testimonials and directions for use.

68 Rush, Benjamin (1745-1813). *Sermons to Gentlemen upon Temperance and Exercise.* Philadelphia: John Dunlap, 1772. Evans: 12547 Austin: 1680 Guerra: a-510. Advice on diet, alcohol use, and exercise. "One of the first American works on personal hygiene" (DAB).

1773

69 Salem (MA) Hospital. *Rules for Regulating Salem Hospital.* [Salem: 1773]. mp: 42498 Bristol: 3637 Austin: 1700 Guerra: a-535. Rules for admission and treatment, along with other hospital regulations.

1774

70 Rush, Benjamin (1745-1813). *An Oration, Delivered February 4, 1774, before the American Philosophical Society, Held at Philadelphia, Containing an Enquiry into the Natural History of Medicine among the Indians of North-America, and a Comparative View of Their Diseases and Remedies, with Those of Civilized Nations.* Philadelphia: Joseph Crukshank, [1774]. Evans: 13592 Austin: 1678 Guerra: a-558. Discussion of Indian health and medicine, including child rearing, diet, treatment of disease, and personal hygiene. Rush compares the Indian lifestyle favorably to the "civilized" nations, suggesting that simplicity leads to health. The work is based entirely on others' experience and, according to Hawke, shows flaws "principally the result of thin material."

1775

71 Jones, John (1729-1791). *Plain, Concise, Practical Remarks on the Treatment of Wounds and Fractures, to Which Is Added a Short Appendix on Camp and Military Hospitals, Principally Designed for the Use of Young Military Surgeons in North America.* New York: John Holt, 1775. Evans: 14134 Austin: 1083 Guerra: a-570. Surgical manual for use by army physicians; it was the first surgical text written in America. Jones bases his discussion on the work of Pott and Le Dran; a 1776 edition included Jones' translation of Von Swieten's "Diseases Incident to Armies." Guerra calls it "the accepted guide to surgical practice" during the Revolution.

1776

72 Morgan, John (1735-1789). *A Recommendation of Inoculation According to Baron Dimsdale's Method.* Boston: J. Gill, 1776. Evans:

14891 Austin: 1336 Guerra: a-590. Recommendation of inoculation for the army with directions; published after British troops re-introduced smallpox to Boston after a long hiatus. Guerra calls it "an elegant and academic work."

1777

73 **Morgan, John (1735-1789).** *A Vindication of His Public Character in the Station of Director-General of the Military Hospitals, and Physician in Chief to the American Army; Anno 1776.* Boston: Powars and Willis, 1777. Evans: 15447 Austin: 1340 Guerra: a-609. Morgan's attempt to counter charges against him which brought about his dismissal as Director of Hospitals during the Revolution. The charges were politically based, caused by those opposed to Morgan's strenuous reforms.

74 **United States Congress.** *In Congress, April 7, 1777. Resolved. . .* [Philadelphia]: John Dunlap, [1777]. Evans: 15660 Austin: 1955 Guerra: a-616. Resolution dealing with the organization of medical hospitals. It establishes a chain of command, the number of directors, doctors, nurses, and general staff. It also discusses duties and rates of pay for each position.

1778

75 **[Brown, William (1752-1792)].** *Pharmacopeia Simplicierum et Efficaciorum, in Usum Nosocomii Militaris, ad Exercitum Foederatarum Americae Civitatum Pertinentis; Hodiernae Nostrae Inopiae Rerumque Angustiis, Feroci Hostium Saevitiae, Belloque Crudeli ex Inopinato Patriae Nostrae Illato Debitis, Maxime Accommodata.* Philadelphia: Styner and Cist, 1778. Evans: 15750 Austin: 297 Guerra: a-636. First pharmacopoeia published in America; designed for military hospitals and based on the *Pharmacopoeia Edinburgensis.* Brown was Physician-General of the Middle Department of the Continental Army, stationed in Lititz, Pennsylvania. This work is sometimes referred to as the "Lititz Pharmacopoeia." In Latin.

76 **[Minor, Jehu (1743-1808)].** *Catechism Physico Medicum, Being an Epitome of the Theory and Practice of Physic; Agreeable to the Best Authors Both Antient and Modern, to Which Is Added Some Practical Rules and Observations and a Discourse on the Nature and Operation of Mercury As a Medicine.* Hartford: Watson and Goodwin, [1778]. Evans: 15913 Austin: 1312 Guerra: a-630. Discussion of nature and practice of "physic" in question and answer form. According to Guerra the references to Boerhaave and Huxham in the preface demonstrate Minor's belief in mercury as a cure-all; the work contains a discussion of mercury as a medicine, particularly for gingivitis, venereal disease, and as a diuretic.

77 **Morgan, John (1735-1789)].** *To the Citizens and Freemen of the United States of America.* Baltimore: M.K. Goddard, [1778]. Evans: 15917 Austin: 1338 Guerra: a-631. Continuation of Morgan's defense of his actions as Director of Hospitals (see 73).

78 **Rush, Benjamin (1745-1813).** *Directions for Preserving the Health of Soldiers: Recommended to the Consideration of the Officers of the Army of the United States.* Lancaster: John Dunlap, 1778. Evans: 16064 Austin: 1633 Guerra: a-635. Suggestions for preventing disease among soldiers; published at the direction of the Board of War. Rush covers diet, dress, hygiene, and other miscellaneous advice. According to Binger "it became a pattern of sound practice in military hygiene."

79 **United States Congress.** *Rules and Directions for the Better Regulating the Military Hospital of the United States: In Consequence of a Resolve of the Honorable the Continental Congress, the 6th of February, 1778; to Be Punctually Observed by the Officers, Nurses, &c. of the Eastern Department, P. Turner, Surg. Gen. M.H.E.D.* Philadelphia: 1778. Evans: 16143 Not included in Readex Austin: 1956 Guerra: a-639. Further regulations for military hospitals.

1780

80 **United States Congress.** *Plan for Conducting the Hospital Department of the United States.* Philadelphia: David C. Claypoole, [1780]. Evans: 17040 Austin: 1951 Guerra: a-680. Reorganization of Hospital Department. The plan lists officers, their duties, and perquisites.

1781

81 **Bayley, Richard, (1745-1801).** *Cases of the Angina Trachealis, with the Mode of Cure, in a Letter to William Hunter, M.D.* New York: H. Gaine, 1781. Evans: 17092 Austin: 160 Guerra: a-681. Description of diphtheria in which Bayley differentiates it from other throat diseases. According to the DAB "by basing his treatment on his sound knowledge of the pathological process, it is said that he cut the mortality rate...nearly in half." Includes a letter from Peter Middleton on croup.

82 **Massachusetts.** *An Act to Incorporate Certain Physicians, by the Name of the Massachusetts Medical Society.* Boston: Benjamin Edes, 1781. Evans: 17228 Austin: 1208 Guerra: a-686. Gives organization and aims of society, including the power to examine physicians and surgeons.

83 **Morgan, John (1735-1789).** *Conclusion of Doctor Morgan's Remarks on Doctor Shippen's Feeble Attempts to Vindicate Himself.* [Philadelphia: David C. Claypoole, 1781]. Evans: 17240 Austin: 1334

Guerra: a-697. Morgan's final charges of misadministration of the Hospital Department; Shippen was Morgan's successor as Director.

84 **[Potter, James (1729-1789)].** *An Oration on the Rise and Progress of Physic in America, Pronounced before the First Medical Society in the Thirteen United States of America Since Their Independence, at Their Convention Held at Sharon, on the Last Day of February, 1780.* Hartford: Hudson and Goodwin, 1781. Evans: 17315 Austin: 1557 Guerra: a-691. Brief history of American medicine, with a review of medical ideas from Hippocrates and Galen to Sydenham, Boerhaave, Pringle, Cadogan, and Tissot.

85 **Rush, Benjamin (1745-1813).** *The New Method of Inoculating for the Small Pox; Delivered in a Lecture in the University of Philadelphia, February 20, 1781.* Philadelphia: Charles Cist, 1781. Evans: 17362 Austin: 1672 Guerra: a-692. Lecture for medical students; includes subjects for inoculation, methods, and cautions. Rush favors the "Suttonian method of inoculation" (Hawke); he refers to cases from Cullen, Brown, and others.

1782
86 **Martin, Hugh (fl. 1780).** *A Narrative of a Discovery of a Sovereign Specific for the Cure of Cancers, with Several Other Improvements Lately Made in Medicine with a Postscript on a Singular Case of a Stone Taken out of the Tongue.* Philadelphia: Robert Aitken, 1782. Austin: 1199 Guerra: a-696. Review of opinions about cancer, with cases Martin claims to have cured (including General Robert Howe). Martin gives no explicit description of his treatment. Martin's "cancer powder" was analyzed by Benjamin Rush in 1786 and found to contain arsenic (a standard ingredient in cancer powders). Hawke suggests it may also have contained marijuana.

87 **Moore, John (1729-1802).** *An Essay on the Causes, Nature, and Cure of Consumptions. In a Letter to a Friend to Which Is Prefixed the Charter of the Massachusetts Medical Society.* Boston: Robert Hodge [1782]. Evans: 17606 Austin: 1329 Guerra: a-713. Discussion of consumption; includes the charter of the Massachusetts Medical Society, as well as a general discussion of the nature of a "medical man."

88 *Observations and Prognostications on the Urine of Human Bodies.* [n.p.: 1782]. mp: 44242 Bristol: 5566 Austin: 1427. Description of urine as diagnostic tool. The pamphlet includes a table of urine conditions and the diseases they indicate.

89 **United States Congress.** *By the United States in Congress Assembled, January 3, 1782.* Philadelphia: David C. Claypoole, 1782. Evans: 17755 Austin: 1948. Deals with organization of military hospitals, including duties of officers, chain of command, means of appointment, and pay.

90 United States Congress. *By the United States in Congress Assembled, July 23rd, 1782.* [Philadelphia: 1782]. mp: 44276 Bristol: 5605 Austin: 1954. Resolution dealing with apothecaries and the General Hospital, regulating accounting procedures and procedures for supplying medicine. It includes miscellaneous resolutions regarding rank for surgeons and allowances for rations and forage.

1784

91 Philadelphia Warden's Office. *Whereas by an Act of General Assembly. . .* [Philadelphia: 1784]. mp: 44592 Bristol: 5962. Regulations preventing ships with infectious sick on board from landing in Philadelphia without a permit from the Health Office.

1785

92 American Academy of Arts and Sciences. *Boston, Nov. 10th, 1785. Sir. . .* Boston: S. Hall, [1785]. Evans: 18901 Austin: 32. Request to physicians and ministers to provide "Bills of Mortality" for towns within Massachusetts. The letter includes a blank table to be filled out with the number of deaths during each month from forty-one different diseases. The Academy proposed to keep records of the "rate of population" and "a natural history of the diseases incident to our climate."

93 Rush, Benjamin (1745-1813). *Observations upon the Cause and Cure of the Tetanus.* [Philadelphia: 1785]. Evans: 19231. Letter to a London physician in which Rush describes his experience in treating tetanus. Rush sees tetanus as a disease of warm climates, leading to a "relaxation"; his cure is the use of "stimulants and tonics" (chiefly large doses of bark and wine) to restore "tone to the system."

1786

94 Humane Society of the Commonwealth of Massachusetts. *Institution of the Humane Society of the Commonwealth of Massachusetts.* [Boston?: 1786?]. mp: 44905 Bristol: 6302 Austin: 997. Rationale for the society and its rules. The pamphlet includes methods for treatment of "persons apparently dead from drowning."

95 [Jackson, Hall (1739-1797)]. *Observations and Remarks on the Putrid Malignant Sore-Throat, Which Has Mortally Raged for Many Years Past.* Portsmouth, NH: John Melcher, 1786. Evans: 19732 Austin: 1032. Description of "throat distemper" epidemic (possibly diphtheria); includes symptoms and treatment.

96 Rush, Benjamin (1745-1813). *Directions for the Use of the Mineral Water and Cold Bath, at Harrogate, Near Philadelphia.* Philadelphia: Melchoir Steiner, 1786. Evans: 19971 Austin: 1634. Extract from Rush's "Observations and Experiments on the Waters of Philadelphia, Abington, and Bristol" (see 758). The pamphlet includes diseases for which mineral waters are effective treatment, both as medicine and as bath. It was published at the request of the owner of the mineral spring at Harrogate.

97 Rush, Benjamin (1745-1813). *An Oration Delivered Before the American Philosophical Society Held in Philadelphia on the 27th of February, 1786, Containing an Enquiry into the Influence of Physical Causes upon the Moral Faculty.* Philadelphia: Charles Cist, 1786. Evans: 19972 Austin: 1676. "Moral faculty" is defined as the ability to distinguish between good and evil. Rush argues that mental competence is based on physical causes—brain "stamina," and heredity among them. Hawke calls it "the major medical paper of his career," Rush's major contribution to the history of psychiatry.

98 Waterhouse, Benjamin (1754-1846). *A Synopsis of a Course of Lectures on the Theory and Practice of Medicine.* Boston: Adams and Nourse, 1786. Evans: 20123 Austin: 2015. Summary of lectures given to "the students of nature in the University of Cambridge." It includes a description of man's place in nature, the causes of disease, chronic diseases, and medical practice.

1787

99 Buchanan, George (1763-1808). *A Treatise upon the Typhus Fever. Published for the Benefit of Establishing a Lying-In Hospital in Baltimore.* Baltimore: William Goddard, 1787. Evans: 21717 Not included in Readex Austin: 346. Apparently a description of typhus; no copy has been found.

100 Mitchill, Samuel Latham (1764-1831). *Observations Anatomical, Physiological, and Pathological on the Absorbent Tubes of the Animal Body.* New York: J. McLean, 1787. Evans: 20527 Austin: 1321. Studies of "lacteal tubes" and "lymphatic tubes" in animals. Mitchill presents examples from lower plant and animal forms; he argues that the same systems exist in humans. The pamphlet includes a second section, "Geological Remarks on Certain Maritime Parts of the State of New York," a report of observations of rocks and clay on the coast of New York.

101 Philadelphia Dispensary. *Plan of the Philadelphia Dispensary for the Medical Relief of the Poor.* [Philadelphia: 1787]. Evans: 19917 Austin: 1513. Proposal for a "public dispensary," a clinic treating the poor without hospitalization. It was the first free dispensary in the country, founded by Benjamin Rush, with Rush and John Jones among the "consulting physicians."

102 *A Treatise on the Gonorrhea.* Norfolk: John McLean, 1787. Evans: 20752 Austin: 1923. Description of history, symptoms, and treatment of gonorrhea.

1788

103 Bard, Samuel (1742-1821). *An Attempt to Explain and Justify the Use of Cold in Uterine Hermorrhagies [sic], with a View to Remove the Prejudices Which Prevail among Women of This City, against the Use of This Safe and Necessary Remedy.* New York: Hugh Gaine, 1788. Evans: 20951 Austin: 115. Argument for the use of cold packs to treat post-partum hemorrhage. Bard cites various authorities and also supports the use of bleeding before delivery. According to the DAB, Bard's favorite branch of medicine was "midwifery."

104 Humane Society of Philadelphia. *Directions for Recovering Persons Who Are Supposed to Be Dead from Drowning, Also for Preventing and Curing the Disorders Produced by Drinking Cold Liquors and by the Action of Noxious Vapours, Lightning, and Excessive Heat and Cold upon the Human Body.* Philadelphia: Joseph James, [1788]. Evans: 21159 Austin: 991. Directions for artificial respiration, both for drowning victims and victims of suffocation by gases. It includes the constitution of the society.

105 Micheau, Paul. *A Dissertation on Hernia Humoralis.* New York: Samuel Campbell, 1788. Evans: 21255 Austin: 1295. Description of hernias, with causes and methods of cure. Micheau disputes the use of opium with hernia.

106 New Haven County (CT) Medical Society. *Cases and Observations; by the Medical Society of New Haven County in the State of Connecticut, Instituted in the Year 1784.* New Haven: Josiah Meigs, 1788. Evans: 21296 Austin: 1258. Twenty-six case descriptions by members of the New Haven Medical Society.

107 Waters, Nicholas Baker (1764-1796). *Tentamen Medicum Inaugurale de Scarlatina Cynanchica.* Philadelphia: William Young, 1788. Evans: 21575 Austin: 2016. Medical dissertation for the University of Pennsylvania on scarlet fever. In Latin.

1789

108 Buchanan, George (1763-1808). *Dissertatio Physiologica Inauguralis de Causis Repirationis Ejusdemque Effectibus.* Philadelphia: Prichard and Hall, 1789. Evans: 21716 Austin: 345. Medical dissertation for the University of Pennsylvania concerning respiration. In Latin.

109 Currie, William (1754-1828). *A Dissertation on the Autumnal Remitting Fever.* Philadelphia: Peter Stewart, 1789. Evans: 21777 Austin: 598. Discussion of symptoms and treatment, accompanied by a discussion of the nervous origin of increased pulse.

110   La Terriere, Pierre de Sales (1747-1834). *A Dissertation on the Puerpal Fever.* Boston: Samuel Hall, 1789. Evans: 21915 Austin: 1124. Medical dissertation for the University at Cambridge (Harvard). It includes the history of the study, symptoms, and competing theories about the nature of the illness.

111   Miller, Edward (1760-1812). *Dissertatio Medica Inauguralis de Physconia Splenica.* Philadelphia: William Young, 1789. Evans: 21964 Austin: 1300. Medical dissertation for the University of Pennsylvania on enlarged spleen. In Latin.

112   Pearson, William. *A Dissertation on the Mixed Fever, Delivered June 30, 1789.* [Boston: 1789]. Evans: 22049 Austin: 1470. Medical dissertation for the University at Cambridge (Harvard). It includes nosology of the disease, diagnosis, case study, and "indications of cure."

113   Rush, Benjamin (1745-1813). *Medical Inquiries and Observations. Vol. I.* Philadelphia: Prichard and Hall, 1789. Evans: 22123 Austin: 1659. Collection of Rush's monographs on various conditions. It is the first statement of Rush's "system of theory and practice," based on the idea that all diseases are due to one "proximate cause": "a state of excessive excitability or spasm in the blood vessels." The volume is "the first larger-than-pamphlet-size collection of medical essays published by an American physician in the United States" (Hawke).

114   Rush, Benjamin (1745-1813). *Observations on the Duties of a Physician and the Methods of Improving Medicine.* Philadelphia: Prichard and Hall, 1789. mp: 45581 Bristol:7053. Austin: 1674. Lecture delivered to medical students at the University of Pennsylvania. It includes advice on conduct and medical practice.

1790
115   Bartram, Moses (d. 1791). *Exercitatio Medica Inauguralis de Victu.* Philadelphia: Thomas Lang, 1790. Evans: 22328 Austin: 150. Medical dissertation for the University of Pennsylvania on nutrition. In Latin.

116   College of Physicians of Philadelphia. *The Charter, Constitution, and Bye Laws of the College of Physicians of Philadelphia.* Philadelphia: Zachariah Poulson, 1790. Evans: 22794 Austin: 495. Organization of the Philadelphia College of Physicians.

117   De Rosset, Armand John (1767-1859). *Dissertatio Medica Inauguralis de Febribus Intermittentibus.* Philadelphia: Thomas Dobson, 1790. Evans: 22460 Austin: 653. Medical dissertation for the University of Pennsylvania on malaria. In Latin.

118 Massachusetts Medical Society. *Medical Papers, Communicated to the Massachusetts Medical Society.* Boston: Thomas and Andrews, 1790. Evans: 22661 Austin: 1221. Collection of medical monographs.

119 Proudfit, Jacob. *Dissertatio Medica Inauguralis de Pleuritide Vera.* Philadelphia: R. Aitken, 1790. Evans: 22823 Austin: 1571. Medical dissertation for the University of Pennsylvania on pleurisy. In Latin.

120 Ramsay, David (1749-1815). *A Dissertation on the Means of Preserving Health in Charleston and the Adjacent Low Country.* Charleston: Markland and McIver, 1790. Evans: 22828 Austin: 1582. Paper read before the South Carolina Medical Society. It is a discussion of preventive medicine, covering infants, children, and adults.

121 Rush, Benjamin (1745-1813). *An Eulogium in Honor of the Late Dr. William Cullen, Professor of the Practice of Physic in the University of Edinburgh.* Philadelphia: Thomas Dobson, 1790. Evans: 22862 Austin: 1641. Eulogy to Cullen as Rush's former teacher. Rush includes a description of Cullen's intellectual achievements, particularly his contributions to chemistry and his revisions of the materia medica.

122 Rush, Benjamin (1745-1813). *An Inquiry into the Effects of Spiritous Liquors on the Human Body.* Boston: Thomas and Andrews, 1790. Evans 22864 Austin: 1656. Description of diseases and other physical disabilities caused by alcohol. Rush also argues that "spiritous liquors" are destructive of life and property (although he exempts cider, beer, and wine from "spirits"). Rush's advocacy of temperance caused him to be formally recognized as the "instaurator" of the American temperance movement (DAB). Rush "recognized addiction to strong drink as a medical and public health problem of the first magnitude" (Binger).

123 Sayre, Francis Bowes (c.1766-1798). *An Inaugural Dissertation on the Causes Which Produce a Predisposition to Phthisis Pulmonalis, and the Method of Obviating Them.* Trenton: Isaac Collins, 1790. Evans: 22874 Austin: 1709. Medical dissertation for the University of Pennsylvania. It includes history and symptoms, a hypothesis for a hereditary disposition, and suggestions for prevention and cure.

124 Waterhouse, Benjamin (1754-1846). *On the Principle of Vitality.* Boston: Thomas and John Fleet, 1790. Evans: 23038 Austin: 2011. Speech to the Massachusetts Humane Society. Waterhouse includes a review of ancient and modern theories regarding the source of human life leading to his conclusion that heat is the source of animation. He also discusses respiration.

1791

125 Bard, John (1716-1799). *A Letter from Dr. John Bard, President of the Medical Society of the State of New York, to the Author of Thoughts on the Dispensary.* New York: 1791. Evans: 23156 Austin: 113. Bard's answer to an earlier attack on the role of the New York Medical Society in the founding for the New York Dispensary. The dispensary was originally founded by Bard in the New York "house of detention"; it became Bellevue Hospital.

126 Barton, William (1754-1817). *Observations on the Progress of Population and the Probability of the Duration of Human Life in the United States of America.* [Philadelphia]: R. Aitken, 1791. Evans: 23158 Austin: 142. Speech to the America Philosophical Society. Barton presents statistics regarding American population in comparison to European statistics. He suggests that Americans have a greater life expectancy and speculates on possible causes. Several tables of data are included.

127 Bayley, Richard (1745-1801). *A Letter from Dr. Richard Bayley to Dr. John Bard, in Answer to a Part of His Letter Addressed to the Author of Thoughts on the Dispensary.* New York: Hugh Gaine, 1791. Evans: 23162 Not included in Readex Austin: 161 Apparently an answer or addition to 125. No copy has been found.

128 Blundell, James. *An Inaugural Dissertation on the Dysentery.* Philadelphia: T. Dobson, 1791. Evans: 23212 Austin: 218. Medical dissertation for the University of Pennsylvania. Contains history, diagnosis, causes, prognosis, and method of cure.

129 Budd, John (1730?-1791). *A Dissertation on Porter.* Charleston: Markland and McIver, 1791. Evans: 23230 Austin: 348. Attack on "porter" imported from London. Budd asserted it was manufactured from polluted Thames water.

130 Conover, Samuel Forman (d. 1824). *An Inaugural Dissertation on Sleep and Dreams, Their Effects on the Faculties of the Mind, and the Causes of Dreams.* [Philadelphia]: T. Lang, 1791. Evans: 23290 Austin: 523. Medical dissertation for the University of Pennsylvania. It is a discussion of the causes and effects of sleep and dreams.

131 Cozens, William R. *An Inaugural Dissertation on the Chemical Properties of Atmospheric Air.* Philadelphia: Thomas Dobson, 1791. Evans: 23296 Austin: 561. Medical dissertation for the University of Pennsylvania. Cozens argues that an understanding of the chemical composition of air is necessary for the preservation of health. He explains the components of air, and includes an extended discussion of hydrogen and "carbonic acid gas" (carbon dioxide).

132 Faugeres, Peter (d. 1798). *A Treatise on Febris Astenica Gravis, or the Severe Asthenic Continued Fever.* New York: Harrisson and Purdy, 1791. Evans: 23364 Austin: 759. Description of the disease, with probable causes and method of cure.

133 Graham, James (fl. 1791). *Disputatio Medica Inauguralis de Scrophula.* Philadelphia: William Young, 1791. Evans: 23419 Austin: 830. Medical dissertation for the University of Pennsylvania on scrofula (tuberculosis of the cervical lymph nodes) In Latin.

134 Handy, Hastings (fl. 1791). *An Inaugural Dissertation on Opium.* Philadelphia: T. Lang, 1791. Evans: 23427 Austin: 875. Medical dissertation for the University of Pennsylvania. It discusses opium production and its effects on the body.

135 Hosack, David (1769-1835). *An Inaugural Dissertation on Cholera Morbus.* New York: Samuel Campbell, 1791. Evans: 23449 Austin: 964. Medical dissertation for the University of Pennsylvania. It discusses the history, cases and method of cure for cholera. Hosack argues against heat as a cause in favor of excess acidity of the stomach.

136 Newman, John (fl. 1791). *A Treatise on Schirrus Tumours and Cancers.* Boston: Nathaniel Coverly, 1791. Evans: 23627. Discussion of tumors and how they become cancerous. Newman claims to be able to cure cancers, but does not detail his treatment. He closes with patient testimonials.

137 New York Dispensary. *Rules of the City Dispensary for the Medical Relief of the Poor.* [New York]: Thomas Greenleaf, [1791]. mp: 46241 Bristol: 7777. Rules and by-laws, including regulations for the apothecaries and for patients.

138 Perkins, Elijah (1767-1806). *An Inaugural Dissertation on Universal Dropsy.* [Philadelphia]: Peter Stewart, 1791. Evans: 23691 Austin: 1492. Medical dissertation for the University of Pennsylvania. Description of the disease with definition, symptoms, causes, case studies, and cure.

139 Pfeiffer, George. *An Inaugural Dissertation on the Gout.* Philadelphia: T. Dobson, 1791. Evans: 23693 Austin: 1499. Medical dissertation for the University of Pennsylvania. Description of the disease with definition and history, causes, and cure.

1792

140 Addoms, Jonas Smith. *An Inaugural Dissertation on the Malignant Fever Which Prevailed in the City of New York during the Months of August, September, and October, in the Year 1791.* New York: T. and J. Swords, 1792. Evans: 24024 Austin: 19. Medical dissertation for Queen's College, New Jersey. It is a description of the yellow fever epidemic in New York, with history, causes, prognosis, and method of cure.

141   Bartlett, John (1756?-1844). *A Discourse on the Subject of Animation.* Boston: Isaiah Thomas and Ebeneezer T. Andrews, 1792. Evans: 24078 Austin: 132. Speech delivered to the Massachusetts Humane Society on resuscitation. Bartlett includes an explanation of circulation and respiration. The publication also includes the Acts of Incorporation for the Humane Society and methods for reviving drowning victims, with case studies.

142   Barton, Benjamin Smith (1766-1815). *An Account of the Most Effectual Means of Preventing the Deleterious Consequences of the Bite of the Crotalus Horridus, or Rattlesnake.* Philadelphia: R. Aitken, 1792. Evans: 142 Not included in Readex Austin: 136. Meisel: 3: p. 357. Apparently Barton's method of treating snakebite; no copy has been found.

143   Bourke, Michael. *A Review on the Subject of Canine Madness.* Philadelphia: Mathew Carey, 1792. Evans: 24138 Austin: 260. Description of hydrophobia, with suggestions for causes and cure.

144   Colesberry, Henry (d. 1849). *Testamen Medicum Inaugurale de Epilepsia.* Wilmington: Brynberg and Andrews, 1792. Evans: 24198 Austin: 93. Medical dissertation for the University of Pennsylvania on epilepsy. In Latin.

145   Currie, William (c.1754 -1828). *An Historical Account of the Climates and Diseases of the United States of America, and of the Remedies and Methods of Treatment Which Have Been Found Most Useful and Efficacious, Particularly in Those Diseases Which Depend upon Climate and Situation.* Philadelphia: T. Dobson, 1792. Evans: 24239 Austin: 600 Hazen: 2732. Study of life expectancy and diseases prevalent in various regions of the country. The work is arranged geographically, beginning with New England. Currie includes some statistics on morbidity, apparently communicated to him by others; he also includes a range of temperatures for each region.

146   Hosack, David (1769-1835). *An Enquiry into the Causes of Suspended Animation from Drowning, with the Means of Restoring Life.* New York: Thomas and James Swords, 1792. Evans: 24409 Austin: 962. Discussion of resuscitation; includes causes of "suspended animation" and treatment.

147   Magruder, Ninian (d. 1823). *Inaugural Dissertation on the Smallpox.* Philadelphia: Zachariah Poulson, 1792. mp: 46495 Bristol: 8051 Austin: 1184. Medical dissertation for the University of

Pennsylvania. Magruder includes a description of the disease and treatment, stressing inoculation and proper treatment by season.

148 Mease, James (1771-1846). *An Inaugural Dissertation on the Disease Produced by the Bite of a Mad Dog, or Other Rabid Animal.* Philadelphia: Thomas Dobson, 1792. Evans: 24534 Austin: 1247. Medical dissertation for the University of Pennsylvania. It includes a description of the disease, definition, symptoms and causes in dogs and humans, and method of cure. Mease had first addressed hydrophobia in an article for *American Magazine* in 1790 while he was still a student. He published a study of quack treatments of hydrophobia in 1808.

149 New Hampshire Medical Society. *Charter of the New Hampshire Medical Society; Together with Their Laws and Regulations.* Exeter: Henry Ranlet, 1792. Evans: 24588 Austin: 1366. Laws, regulations, act of incorporation and list of fellows for the New Hampshire Medical Society.

150 Seaman, Valentine (1770-1817). *An Inaugural Dissertation on Opium.* Philadelphia: Johnson and Justice, 1792. Evans: 24775 Austin: 1720. Medical dissertation for the University of Pennsylvania. It includes a description of opium's effects, its mode of operation, and its uses.

151 Thomas, Tristram (1769-1847). *Disputatio Medica Inauguralis, de Pneumonia Sthenica.* Philadelphia: William Young, 1792. Evans: 24848 Austin: 1895. Medical dissertation for the University of Pennsylvania on pneumonia. In Latin.

152 *To the Inhabitants of South Carolina.* [Charleston: 1792]. mp: 46587 Bristol: 8154 Austin: 1800. Proposal for setting physicians' fees; includes a sample fee table.

153 United States Treasury Department. *Report.* [Philadelphia: 1792]. Evans: 24933. Report on the advisability of establishing marine hospitals, with suggestions for funding.

154 Van Solingen, Henry. *An Inaugural Dissertation on Worms of the Human Intestine.* New York: T. and J. Swords, 1792. Evans: 24953 Austin: 1973. Medical dissertation for Queen's College, New Jersey. It includes a description of various types of worms, a discussion of their "origin and nourishment," causes, symptoms, diagnosis, and method of cure.

155 Waterhouse, Benjamin (1754-1846). *The Rise, Progress, and Present State of Medicine.* Boston: Thomas and John Fleet, 1792. Evans: 24987 Austin: 2014. Paper presented before the Middlesex (Massachusetts) Medical Association. It shows an "emphasis on experimental investigation [which] reveals Waterhouse as a man far in advance of his time" (DAB). Waterhouse was professor of the theory and practice of medicine at Harvard Medical School.

156   Woodhouse, James (1770-1809). *An Inaugural Dissertation on the Chemical and Medical Properties of the Persimmon Tree and the Analysis of Astringent Vegetables.* Philadelphia: William Woodhouse, 1792. Evans: 25055 Austin: 2086. Medical dissertation for the University of Pennsylvania on the persimmon tree; includes chemical analysis with various reagents, a discussion of the analytical results, instructions for the "pharmaceutical treatment" of the persimmon and suggestions for medical and commercial use. The dissertation met with considerable acclaim and may have influenced Woodhouse to move from medicine to chemistry (he founded the Chemical Society of Philadelphia in 1792).

1793

157 *An Account of the Rise, Progress, and Termination of the Malignant Fever, Lately Prevalent in Philadelphia.* Philadelphia: Benjamin Johnson, 1793. Evans: 25075 Austin: 15. Narrative of the yellow fever epidemic including lists of burials and some medical accounts. It supports the theory that the disease was imported and contagious. It concludes with "A Short Account of the Plague in London, 1665."

158   Andrews, John (fl. 1793). *An Inaugural Dissertation on the Apoplexy.* Philadelphia: Parry Hall, 1793. Evans: 25111 Austin: 46. Medical dissertation for the University of Pennsylvania. It includes definition, diagnosis, causes, prognosis, and cure, along with appearance on dissection.

159   Borrowe, Samuel (c.1766-1828). *An Inaugural Dissertation on the Cynanche Trachealis.* New York: T. and J. Swords, 1793. Evans: 25213 Austin: 221. Medical dissertation for Columbia. It includes a history, diagnosis, causes, prognosis, symptoms, cure, and case studies.

160   [Burdon, William.] *The Gentleman's Pocket-Farrier.* Springfield, MA: James Hutchins, 1793. Evans: 25238 Austin: 352. Instructions for buying and caring for horses; includes simple veterinary treatment.

161   Buxton, Charles (1768-1833). *An Inaugural Dissertation on the Measles.* New York: T. and J. Swords, 1793. Evans: 25249 Austin: 371. Medical dissertation for Queen's College, New Jersey. It includes definition and history, causes, diagnosis, prognosis, general observations, appearances on dissection, method of cure, and case studies.

162   Carey, Mathew (1760-1839). *A Desultory Account of the Yellow Fever Prevalent in Philadelphia and of the Present State of That*

City. [Philadelphia: Mathew Carey, 1793]. mp: 46709 Bristol: 8300
Austin: 401. A narrative of the 1793 yellow fever epidemic.

163 Carey, Mathew (1760-1839). *Observations on Dr. Rush's
Enquiry into the Origin of the Late Epidemic Fever in Philadelphia.*
Philadelphia: Mathew Carey, 1793. Evans: 25254 Austin: 406. Attack
on Rush's theories regarding the cause of yellow fever (see 181).
Carey argues strongly that the disease was brought into Philadelphia by
ships returning from the Caribbean.

164 Carey, Mathew (1760-1839). *Short Account of the Malignant
Fever Lately Prevalent in Philadelphia with a Statement of the
Proceedings That Took Place on the Subject in Different Parts of the
United States.* Philadelphia: Mathew Carey, 1793. Evans: 25257
Austin: 408. Carey's famous history of the yellow fever epidemic.
Includes medical information, burial statistics, and meteorological
information from David Rittenhouse. The work went through eight
editions and was translated into French and German.

165 Clarke, John (1755-1798). *A Discourse, Delivered before the
Humane Society of the Commonwealth of Massachusetts, at the Semi-
Annual Meeting, Eleventh of June, 1793.* Boston: Belknap and Hall,
1793. Evans: 25303 Austin: 476. General discussion of effects
produced on the body by drowning, with an explanation of the actions
recommended by the Humane Society to counteract these effects.
Clarke was minister of First Church in Boston.

166 Columbia College, New York. *An Ordinance for Conferring
the Degree of Doctor of Medicine in Columbia College.* [New York]: T.
Greenleaf, [1793]. Evans: 25315 Austin: 506. Rules for admission to
the doctoral program and examination and dissertation requirements.

167 Cornelison, Abraham. *Inaugural Dissertation on the Pertussis or
Whooping Cough.* New York: T. and J. Swords, 1793. Evans: 25350
Austin: 546. Medical dissertation for Queen's College, New Jersey. It
includes history, diagnosis, causes, prognosis, and cure.

168 Currie, William (c.1754-1829). *A Description of the
Malignant, Infectious Fever Prevailing at Present in Philadelphia.*
Philadelphia: T. Dobson, 1793. Evans: 25366 Austin: 597.
Description of yellow fever; includes description of disease, symptoms,
hypothesis about causes, treatment, and cure. Currie was a strong
believer in the contagious nature of the disease.

169 Hicks, John B. *An Inaugural Dissertation on Compression of the
Brain from Concussion.* New York: T. and J. Swords, 1793. Evans:
25602 Austin: 901. Medical dissertation for Columbia University.
Includes description, causes, predisposition, and cure.

170 Hirte, Tobias. *Dr. Van Swieten's, Late Physician to His Imperial Majesty, Renowned Pills.* [Philadelphia: 1793]. mp: 46779 Bristol: 8365 Austin: 914. Advertisement for patent medicine.

171 Johnson, Thomas (1760-1831). *An Inaugural Dissertation on the Putrid Ulcerous Sore Throat.* Philadelphia: Parry Hall, 1793. Evans: 25663 Austin: 1077. Medical dissertation for the University of Pennsylvania; includes description, symptoms, causes, diagnosis, prognosis, management, and cure.

172 Johnston, Robert (d. 1808). *An Inaugural Dissertation on the Influenza.* Philadelphia: Johnson and Justice, 1793. Evans: 25664 Austin: 1078. Medical dissertation for the University of Pennsylvania; includes definition, history, diagnosis, causes, cure, prophylactics, and prognosis.

173 Medical Society in the State of Connecticut. *The Act Incorporating the Medical Society in the State of Connecticut.* New Haven: T. and S. Green, [1793]. mp: 46820 Bristol: 8411. Act of incorporation with constitution and by-laws.

174 Nassy, David de Isaac Cohen. *Observations on the Cause, Nature and Treatment of the Epidemic Disorder Prevalent in Philadelphia.* Philadelphia: Parker & Co., 1793. Evans: 25854 Austin: 1358. Description of the yellow fever epidemic. In French, with English translation on the facing pages. Nassy was a member of the American Philosophical Society.

175 Newnan, John (fl. 1793). *An Inaugural Dissertation on General Dropsy.* Philadelphia: Parry Hall, 1793. Evans: 25922 Austin: 1373. Medical dissertation for the University of Pennsylvania. Includes definition, history and symptoms, diagnosis, causes, prognosis, and cure.

176 New York Committee to Prevent the Introduction and Spreading of Infectious Diseases. *As It Is a Point Agreed on by All Writers...* [New York: 1793]. mp: 46832 Bristol: 8426 Austin: 1376. Measures to prevent the spread of yellow fever from Philadelphia to New York, including a quarantine on an goods from Philadelphia.

177 Philadelphia Medical Society. *The Act of Incorporation and Laws of the Philadelphia Medical Society.* Philadelphia: Johnston and Justice, 1793. Evans: 25996 Austin: 1518. Includes constitution and by-laws.

178 Pope, John. *Cancers.* [Boston: 1793]. Evans: 26021 Austin: 1550. Chiefly patient testimonials regarding Pope's treatment of cancer.

179 Post, Jotham (1771-1817). *An Inaugural Dissertation, to Disprove the Existence of Muscular Fibres in the Vessels.* New York: T. and J. Swords, 1793. Evans: 26028 Austin: 1555. Medical dissertation for Columbia University. Post argues against the existence of muscles in the blood vessels and in favor of the heart as the sole source of the motion of the blood.

180 Rush, Benjamin (1745-1813). *[An Account of the Causes of Longevity.* Philadelphia: 1793.]* mp: 46871 Not in Readex Bristol: 8470. No copy is known to exist.

181 Rush, Benjamin (1745-1813). *An Enquiry into the Origin of the Late Epidemic Fever in Philadelphia, in a Letter to Dr. John Redman, President of the College of Physicians.* Philadelphia: Mathew Carey, 1793. Evans: 26111 Austin: 1639. Rush's hypothesis for the cause of yellow fever in "vegetable and animal putrefaction." This was a controversial theory since it assumed environmental causes for yellow fever in Philadelphia.

182 Rush, Benjamin (1745-1813). *Medical Inquiries and Observations, Volume II.* Philadelphia: T. Dobson, 1793. Evans: 26112 Austin: 1659. Collection of Rush's medical monographs on a variety of diseases and conditions including pulmonary consumption, measles, "bilious and remitting fevers," influenza, dropsy, and the diseases of old age.

183 Sawyer, Matthias Enoch. *An Inaugural Dissertation, Containing an Inquiry into the Living Principles and Causes of Animal Life.* Philadelphia: T. Dobson, 1793. Evans: 26140 Austin: 1707. Medical dissertation for the University of Pennsylvania.

184 Seybert, Adam (1773-1825). *An Inaugural Dissertation, Being an Attempt to Disprove the Doctrine of the Putrefaction of the Blood of Living Animals.* Philadelphia: T. Dobson, 1793. Evans: 26153 Austin: 1734 Meisel: 3:358. Medical dissertation for the University of Pennsylvania. Seybert attacked the theory that some diseases cause "putrefaction" of the blood (asserted by Boerhaave, among others). His argument was based upon animal experiments. His dissertation was reprinted in 1805 in a collection of outstanding theses from American medical schools (DAB).

185 Stokes, William. *Tentamen Medicum Inaugurale, Quaedam de Asphyxia, ab Aeres Dephlogisticati, Privatione Oriunda, Tradens.* Philadelphia: William Young, 1793. Evans: 26214 Austin: 1830. Medical dissertation for the University of Pennsylvania on suffocation ("dephlogisticated air" was Priestley's term for oxygen). In Latin.

186 Taylor, Willet, Jr. (d. 1811). *An Inaugural Dissertation on the Scarlatina Anginosa, as It Prevailed in This City.* New York: T. and J.

Swords, 1793. Evans: 26246 Austin: 1866. Medical dissertation for Columbia University; includes history, description, causes, prognosis, and cure for scarlet fever.

187 **Thomson, J.** *Modern Practice of Farriery, or Complete Horse Doctor.* New York: Berry, Rogers, and Berry, [1793]. Bristol: 8490 Not included in Readex. A veterinary manual.

188 **Wallace, James Westwood.** *An Inaugural Physiological Dissertation on the Catamenia, to Which are Subjoined Observations on Amenorrhea.* Philadelphia: 1793. Evans: 26414 Austin: 1989. Medical dissertation for the University of Pennsylvania on menstruation.

189 **Wilkins, Henry (1767-1847).** *The Family Adviser, or a Plain and Modern Practice of Physic; Calculated for the Use of Private Families and Accommodated to the Diseases of America.* Philadelphia: John Dickins, 1793. Evans: 26482 Austin: 2046. Home medical guide with descriptions of diseases, causes, and treatment. It includes a copy of John Wesley's *Primitive Physic.*

190 **Wilkins, Henry (1767-1847).** *An Inaugural Dissertation on the Theory and Practice of Emetics.* Philadelphia: Parry Hall, 1793. Evans: 26483 Austin: 2053. Medical dissertation for the University of Pennsylvania; covers theory and practice, including a list of diseases which are treated with emetics.

191 **Williamson, Matthias Hampton.** *An Inaugural Dissertation on the Scarlet Fever, Attended with an Ulcerated Sore Throat.* Philadelphia: Johnston & Justice, 1793. Evans: 26489 Austin: 2065. Medical dissertation for the University of Pennsylvania; includes definition, history, causes, and cure.

192 **Youle, Joseph (d. 1795).** *An Inaugural Dissertation on Respiration: Being an Application of the Principles of the New Chemistry to That Function.* New York: T. and J. Swords, 1793. Evans: 26520 Austin: 2102. Medical dissertation for Columbia University. It is a chemical analysis of respiration, including a description of the gases involved and their effect on the body.

1794

193 **Abeel, David G. (d.1795).** *Inaugural Dissertation on Dysentery.* New York: T. and J. Swords, 1794. Evans: 26530 Austin: 1. Medical dissertation for Columbia University; includes description, history, dissections, diagnosis, causes, prognosis, and methods of cure.

194 **[Bordley, John Beale (1727-1804)].** *Yellow Fever.* [Philadelphia: Charles Cist?, 1794]. Evans: 26683 Austin: 220.

Pamphlet on yellow fever distributed to members of Congress (then meeting in Philadelphia) to prevent panic. Bordley compares yellow fever to smallpox and argues that yellow fever is conveyed by immigrants and only during the summer.

195 Carey, Mathew (1760-1839). *Address of M. Carey to the Public.* [Philadelphia: Mathew Carey, 1794]. Evans: 26729 Austin: 400. Carey's answer to charges concerning his conduct while he was a member of the Philadelphia Committee of Health during the yellow fever epidemic.

196 Carey, Mathew (1760-1839). *Gentlemen. . . .* [Philadelphia: Mathew Carey, 1794]. Evans: 26731 Austin: 407. Carey's argument that yellow fever was reappearing during September, 1794. This pamphlet is addressed to the Philadelphia Committee of Health; Carey argues that the committee should head off a panic that would be ruinous to business.

197 *Catalogue of Books Belonging to the Medical Library in the Pennsylvania Hospital to Which Are Prefixed the Rules to Be Observed in the Use of Them.* Philadelphia: Zachariah Poulson, 1794. Evans: 27510 Austin: 1532. List of books and procedures.

198 Cathrall, Isaac (1764-1819). *A Medical Sketch of the Synochus Maligna, or Malignant Contagious Fever As It Lately Has Appeared in the City of Philadelphia.* Philadelphia: T. Dobson, 1794. Evans: 26747 Austin: 432. Clinical description of yellow fever. Cathrall begins with a synopsis of the epidemic, then gives a definition, description and method of cure for the disease. He closes with observations based on dissection.

199 *Charter for Establishing an Hospital in the City of New York.* New York: Hugh Gaine, 1794-97. Evans: 27406 Austin: 1796. Description of officers of the New York Hospital with rules governing elections and procedures.

200 Cheever, Abijah (1760-1843). *History of a Case of Incisted Dropsy with a Dissection of Several Cysts.* [Boston: 1794]. Evans: 26763 Austin: 451. Case history of dropsy patient; first delivered as an address to the American Academy of Arts and Sciences. Cheever includes an autopsy account and closes with two drawings of the contents of the cysts.

201 [Clark, Thaddeus]. *An Account of a Remarkable Case of Tetanus.* Norwich: Thomas Hubbard, 1794. Evans: 26535 No copy in Readex Austin: 471. No copy has been found.

202 Condict, Lewis (1772-1862). *Inaugural Dissertation on the Effects of Contagion upon the Human Body.* Philadelphia: William W.

Woodward, 1794. Evans: 26802 Austin: 514. Medical dissertation for the University of Pennsylvania. Condict concentrates particularly on the "febrile contagious diseases."

203 Coxe, John Redman (1773-1864). *An Inaugural Dissertation on Inflammation.* Philadelphia: R. Aitken and Son, 1794. Evans: 26828 Austin: 556. Medical dissertation for the University of Pennsylvania, concentrating on the blood and showing a relation between blood and heat. Coxe went on to become the "father of pharmacy in Philadelphia" and helped James Woodhouse found the Chemical Society of Philadelphia (DAMB).

204 Currie, William (c.1754-1828). *An Impartial Review of That Part of Dr. Rush's Late Publication Entitled "An Account of the Bilious Remitting Yellow Fever, As It Appeared in the City of Philadelphia in the Year 1793," Which Treats of the Origin of the Disease.* Philadelphia: Thomas Dobson, 1794. Evans: 26836 Austin: 601. Currie's attack on Rush's theories concerning the origin of yellow fever. Currie asserts that yellow fever is brought into the city by outsiders; he insists upon the "wholesomeness" of Philadelphia.

205 Currie, William (1754-1828). *A Treatise on the Synochus Icteroides or Yellow Fever As It Lately Appeared in the City of Philadelphia.* Philadelphia: Thomas Dobson, 1794. Evans: 26837 Austin: 607. Clinical discussion of the yellow fever epidemic; includes definition, origin, methods of treatment, and method of prevention.

206 Cutbush, Edward (1772-1843). *An Inaugural Dissertation on Insanity.* Philadelphia: Zachariah Poulson, 1794. Evans: 26838 Austin: 611. Medical dissertation for the University of Pennsylvania. Cutbush treats insanity as an illness.

207 Davidson, Robert Gamble Waiters (1779-1804). *An Inaugural Dissertation on the Suffocatio Stridula, or Croup.* Philadelphia: Wrigley & Berriman, 1794. Evans: 26852 Austin: 627. Medical dissertation for University of Pennsylvania; includes definition, causes, and treatment.

208 Deveze, Jean (1753-1829). *An Enquiry into and Observations upon the Causes and Effects of the Epidemic Disease Which Raged in Philadelphia from the Month of August till towards the Middle of December, 1793.* Philadelphia: Parent, 1794. Evans: 26873 Austin: 658. Discussion of yellow fever epidemic; French with English translation on the facing pages. Deveze, "the principal physician of the military hospital established by the French Republic at Philadelphia," disputed the idea that yellow fever was contagious. He based his conclusion on the fact that the doctors, nurses, and attendants at the Bush Hill hospital who attended yellow fever victims did not contract the disease.

209 Dingley, Amasa (d. 1798). *An Oration on the Improvement of Medicine.* New York: John Buel, 1794. Evans: 26892 Austin: 671. Speech delivered to the New York Medical Society recommending research and publication in medicine.

210 Drysdale, Thomas (1770-1798). *Tentamen Medicum Inaugurale Varia de Hepat Proferens.* Philadelphia: Thomas Dobson, 1794. Evans: 26914 Austin: 701. Medical dissertation for the University of Pennsylvania on the liver. In Latin.

211 Irving, Peter (1771-1838). *An Inaugural Dissertation on the Influenza.* New York: T. and J. Swords, 1794. Evans: 27158 Austin: 1030. Medical dissertation for Columbia University. Irving was Washington Irving's brother and is better known as a literary figure and political writer than as a physician.

212 [Jones, Absalom (1746-1818) and Richard Allen]. *A Narrative of the Proceedings of the Black People during the Late Awful Calamity in Philadelphia in the Year 1793.* Philadelphia: William W. Woodward, 1794. Evans: 27170 Austin: 1079. Description of public service performed by Philadelphia's black population during the yellow fever epidemic; refutation of Mathew Carey's charges of black profiteering.

213 Jones, Calvin (1775-1846). *A Treatise on the Scarlatina Anginosa or What Is Vulgarly Called the Scarlet Fever or Canker Rash.* Catskill: M. Croswell & Co., 1794. Evans: 27171 Austin: 1080. Description of scarlet fever; includes method of treatment and conjectures concerning causes.

214 Lamb, John Jr. *An Inaugural Dissertation on the Apoplexy.* Philadelphia: Jones, Hoff & Daniels, 1794. Evans: 27193 Austin: 1118. Medical dissertation for the University of Pennsylvania; includes description, history, causes, and method of treatment.

215 Ludlow, Edmund. *An Inaugural Dissertation on Intermittent Fever.* New York: T. and J. Swords, 1794. Evans: 27238 Austin: 1165. Medical dissertation for Columbia; includes description, history, causes, and treatment.

216 Mead, Henry (d. 1838). *An Inaugural Dissertation on the Cholera Morbus.* New York: T. and J. Swords, 1794. Evans: 27303 Austin: 1244. Medical dissertation for Columbia; includes history and definition, causes, prognosis, and method of cure.

217 *Minutes of the Proceedings of the Committee Appointed on the 14th September, 1793, by the Citizens of Philadelphia, the Northern Liberties, and the District of Southwark to Attend to and Alleviate the*

Sufferings of the Afflicted with the Malignant Fever Prevalent in the City and Its Vicinity. Philadelphia: R. Aitken and Son, 1794. Evans: 27501 Austin: 1509. Accounts of general meetings through 1793 with reports from physicians and others. It includes an appendix with the names of those hospitalized and their disposition (i.e. dead, convalescing, or discharged).

218 **Rose, Henry (fl. 1794).** *An Inaugural Dissertation on the Effects of the Passions upon the Body*. Philadelphia: William W. Woodward, 1794. Evans: 27638 Austin: 1622. Medical dissertation for the University of Pennsylvania; on physical effects of emotions.

219 **Rush, Benjamin (1745-1813).** *An Account of the Bilious Remitting Yellow Fever As It Appeared in the City of Philadelphia in the Year 1793*. Philadelphia: Thomas Dobson, 1794. Evans: 27658 Austin: 1631. Rush's account of the yellow fever epidemic. The pamphlet describes the history of the Philadelphia experience, the history and symptoms of the disease, and speculates about causes and treatment (along with causes and treatments used by others, which he dismisses). Winslow calls it Rush's "first major contribution to epidemiology." Rush believed the fever resulted from local causes rather than being imported, but he also believed the disease was contagious.

220 **Taplin, William (d. 1807).** *The Gentleman's Stable Directory or Modern System of Farriery*. 12th ed. Philadelphia: T. Dobson, 1794. Evans: 27771 Austin: 1858. Veterinary guide for horses along with some advice for buying and caring for them. It also includes treatment for distemper in dogs.

1795

221 **Alexander, Ashton.** *An Inaugural Dissertation on the Influence of One Disease in the Cure of Others*. Philadelphia: Alexander McKenzie, 1795. Evans: 219 Austin: 26. Medical dissertation for the University of Pennsylvania; concerned with interaction among diseases.

222 **Anderson, Peter S. (d. 1803).** *An Inaugural Dissertation on the Diarrhoea Infantum*. New York: Tiebout & O'Brien, 1795. Evans: 28161 Austin: 43. Medical dissertation for Columbia.

223 **Brooks, John (1752-1825).** *A Discourse Delivered before the Humane Society of the Commonwealth of Massachusetts*. Boston: T. Fleet, 1795. Evans: 28351 Austin: 278. General history of science and medicine as "systems of philosophy."

224 *The Cheap and Famous Farrier*. Ephrata: 1795. mp: 47381 Bristol: 9053. Guide for treatment of diseases and wounds in horses.

225 Clark, Thaddeus. *A Treatise on the Scarlatina Anginosa with an Appendix Containing Observations on the Practice with Salt and Vinegar.* Norwich: T. Hubbard, 1795. Evans: 28421 Austin: 472. Description of the disease with treatment; appendix describes vinegar and salt treatment, which was apparently widespread.

226 Davis, Matthew Livingston (c.1773-1850). *A Brief Account of the Epidemical Fever Which Lately Prevailed in the City of New York.* New York: Matthew L. Davis, 1795. Evans: 28538 Austin: 629. Discussion of yellow fever, including a collection of reports from various sources. An appendix includes a casualty list. Davis was the editor of the *Evening Post.*

227 Dix, William (d. 1799). *An Inaugural Dissertation on the Dropsy.* Worcester: Isaiah Thomas, Jr., 1795. Evans: 28572 Austin: 676. Medical dissertation for Harvard; includes description, history, symptoms, causes, treatment, and method of cure.

228 Everett, Charles. *An Inaugural Dissertation on the Function of Menstruation.* Philadelphia: S. H. Smith, 1795. Evans: 28645 Austin: 738. Medical dissertation for the University of Pennsylvania.

229 Fleet, John, Jr. (1766-1813). *Dissertatio Inauguralis Medica, Sistens Observationes ad Chirurgiae Operationes Pertinentes.* Boston: Thomas Fleet, 1795. Evans: 28679 Austin: 778. Medical dissertation on surgical practice. In Latin.

230 Hedges, Phinehas. *Strictures on the Elementa Medicinae of Doctor Brown.* Goshen: David M. Westcott, 1795. Evans: 28816 Austin: 896. Critical discussion of John Brown; attack on his work and theories.

231 Humane Society of the State of New York. *The Constitution of the Humane Society of the State of New York.* New York: J. Buel, 1795. Evans: 29202 Austin: 1000. Constitution of Humane Society; includes "Address to the Citizens," directions for artificial respiration.

232 Jewett, Paul. *The New England Farrier, or a Compendium of Farriery in Four Parts.* Newburyport: William Barrett, 1795. Evans: 28901 Austin: 1061. Veterinary manual covering diseases of horses, cattle, sheep, and pigs, with suggested remedies.

233 Jones, John (1729-1791). *The Surgical Works of the Late John Jones, M.D. to Which Are Added a Short Account of the Life of the Author with Occasional Notes and Observations by James Mease, M.D.* Philadelphia: Wrigley and Berriman, 1795. Evans: 28909 Austin: 1086. Collection of Jones' works including "Plain, Concise, Practical Remarks on the Treatment of Wounds and Fractures" (see 71), as well as "A Case of Anthrax" and "An Uncommon Case of Hydrocele." The work includes biographical material and notes by James Mease.

234 Lee, Samuel Holden Parsons (1771-1863). *Medical Advice to Seamen with Directions for a Medicine Chest.* New London: Samuel Green, 1795. mp: 47482 Bristol: 9168 Austin: 1137. Describes four basic types of disease (fever, inflammation, flux, and weakness) with treatment. The pamphlet also includes lists of medicines recommended for a sea chest.

235 *Letter from Doctors Taylor and Hanford.* [Norfolk?: 1795?]. mp: 47483 Not included in Readex Bristol: 9169. According to Bristol, concerns fever in Norfolk in 1795.

236 May, Frederick (1773-1847). *An Inaugural Dissertation on the Animating Principle or Anima Mundi, How Afforded and How Acting in Man, and How Acted upon in That Disease Commonly Denominated Tetanus or Lock Jaw.* Boston: William Spotswood, 1795. Evans: 29056 Austin: 1242. Medical dissertation for Harvard; chiefly on tetanus.

237 Mitchill, Samuel Latham (1764-1831). *Remarks on the Gaseous Oxyd of Azote or of Nitrogene and on the Effects It Produces When Generated in the Stomach, Inhaled into the Lungs, and Applied to the Skin.* New York: T. and J. Swords, 1795. Evans: 29089 Austin: 1323. Mitchill's theory that a combination of oxygen and "nitrous acid" (Priestly's "dephlogisticated nitrous air") is the cause of disease. According to the DAB, "his theory of the septic substance he called 'septon,' though fanciful and erroneous, was an incentive to the study of sanitary chemistry and hygiene and was one of the factors that led Davy to investigate problems in nitrous oxide." Mitchill was professor of chemistry at the College of New York.

238 New York (City) Committee of Health. *Names of Persons Who Have Died in New York of the Yellow Fever from the 29th of July to the Beginning of November, 1795.* New York: 1795. Evans: 29196 Austin: 1387. Begins with an address to the public, outlining the situation regarding the epidemic and the counter-measures which have been taken. The names of the dead are listed, sometimes with their occupations.

239 New York (City) Dispensary. *Rules for the City Dispensary for the Medical Relief of the Poor.* New York: Thomas Greenleaf, [1795]. Evans: 29201 Austin: 1385. Rules and by-laws of the charity hospital.

240 Philadelphia Board of Health. *Health-Office, Port of Philadelphia, June 30, 1795.* [Philadelphia: 1795]. Evans: 29304

Austin: 1504. Rules governing the landing of ships with diseased crew or passengers; another response to the yellow fever epidemic.

241 **Ross, William Morrey (d. 1818).** *A Chemico-Physiological Inaugural Dissertation on Carbone or Charcoal.* New York: T. and J. Swords, 1795. Evans: 29431 Austin: 1624. Medical dissertation for Columbia; description of charcoal and its uses, closing with its medical uses.

242 **Shultz, Benjamin.** *An Inaugural Botanico-Medical Dissertation on the Phytolacca Decandra of Linnaeus.* Philadelphia: Thomas Dobson, 1795. Evans: 29510 Austin: 1746 Meisel: 3:359. Medical dissertation for the University of Pennsylvania; description of Phytolacca Decandra with its medical uses. The dissertation includes a plate and figures.

243 **Spaulding, Mary (b. 1769).** *Remarkable Narrative of Mary Spaulding, Daughter of Benjamin Spaulding of Chelmsford.* Boston: Manning and Loring, 1795. Evans: 29554 Austin: 1810. Autobiography of an invalid; an extended case study.

244 **Stuart, James.** *Directions for Medicine Chests.* Philadelphia: Ormrod and Conrad, 1795. mp: 47615 Not included in Readex Bristol: 9312 Austin: 112. Apparently list of medicines for sea chests.

245 **Webster, Noah (1758-1843).** *To the Physicians of Philadelphia, New York, Baltimore, Norfolk, and New Haven.* New York: 1795. mp: 47676 Bristol: 9417 Austin: 2024. Circular letter requesting information on various aspects of yellow fever, particularly about its origin. Webster proposed to publish a collection of material (see 268).

246 **Wetmore, Timothy Fletcher (1764-1799).** *An Inaugural Dissertation on Puerpal Fever.* New York: T. and J. Swords, 1795. Evans: 29881 Austin: 2033. Medical dissertation for Columbia; includes description, causes, and treatment.

1796

247 **Allen, Israel.** *A Treatise on the Scarlatina Anginosa and Dysentery, and Sketches on Febrile Spasm As Produced by Phlogiston.* Leominster, MA: Charles Prentiss, 1796. Evans: 29966 Austin: 29. Description of both diseases with history, symptoms, causes, and treatment based on the epidemics of 1786. This discussion is followed by a review of research results regarding the influence of phlogiston on fever.

248 **Anderson, Alexander (1775-1870).** *An Inaugural Dissertation on Chronic Mania.* New York: T. and J. Swords, 1796. Evans: 29990 Austin: 40. Medical dissertation for Columbia. Anderson later achieved

fame as an engraver after he gave up his medical practice. Some of his
engravings (e.g. those for *Albinus' Anatomy* and for Berwick's *General
History of Quadrupeds*) have scientific or medical subject matter.

249 Ball, Thomas. *An Inaugural Dissertation on the Causes and
Effects of Sleep.* Philadelphia: Budd and Bartram, 1796. Evans: 30014
Austin: 104. Medical dissertation for the University of Pennsylvania;
includes some consideration of sleep disorders.

250 Bayley, Richard (1745-1801). *An Account of the Epidemic
Fever Which Prevailed in the City of New York during Part of the
Summer and Fall of 1795.* New York: T. and J. Swords, 1796.
Evans: 30041 Austin: 159. Description of the yellow fever epidemic
with suggestions for the most probable causes and most effective
treatment. It is based on Bayley's observations as a health officer (he
was physician to the Port of New York); he stresses the seasonal nature
of the disease and argues that it is contagious rather than infectious.

251 Caldwell, Charles (1772-1853). *An Attempt to Establish the
Original Sameness of Three Phenomena of Fever (Principally Confined
to Infants and Children) Described by Medical Writers under the Several
Names of Hydrocephalus Internus, Cynanche Trachealis, and Diarrhea
Infantum.* Philadelphia: Thomas Dobson, 1796. Evans: 30148 Austin:
385. A medical dissertation for the University of Pennsylvania.
Caldwell tries to differentiate among three types of childhood diseases.

252 [Carter, George (fl. 1796)]. *An Essay on Fevers, Particularly
on the Fever Lately So Rife in Charleston, South Carolina.* Charleston:
W.P. Harrison and Co., 1796. Evans: 20167 Austin: 419. Chiefly
concerned with yellow fever; supports Sydenham's theory of blood
circulation disorders as the source of disease.

253 Chisholm, Robert (fl. 1796). *An Inaugural Dissertation on the
Hydrocephalus Internus or Internal Dropsy of the Brain.* Philadelphia:
Ormrod and Conrad, 1796. Evans: 30191 Austin: 461. Medical
dissertation for the University of Pennsylvania; includes history,
diagnosis, symptoms, prognosis, causes, and cure.

254 Humane Society of the Commonwealth of Massachusetts.
*Summary of the Method of Treatment to Be Used with Persons
Apparently Dead from Drowning.* [Boston: 1796?]. mp: 47811
Bristol: 9579. Broadside giving directions for a series of steps in
artificial respiration.

255 Jones, Edward (fl. 1796). *An Inaugural Dissertation on
Pneumonia or Pulmonary State of Fever.* Philadelphia: Ormrod and
Conrad, 1796. Evans: 30643 Austin: 1081. Medical dissertation for
the University of Pennsylvania; includes description, causes, and
treatment.

256 [**Jones, Ira (fl. 1796)**]. *A New Treatise on the Consumption.* Newfield, CT: Ira Jones, 1796. Evans: 30645 Austin: 1082. Discussion of disease with some apparently controversial suggestions for treatment; contains attack on Rush.

257 **Lee, Samuel (fl. 1796).** *Lee's Genuine Windham Bilious Pills.* [n.p.: 1796?]. mp: 47821 Bristol: 9591 Austin: 1135. Advertisement for patent medicine.

258 **Mackrill, Joseph (1762-1820).** *The History of the Yellow Fever, with the Most Successful Method of Treatment.* Baltimore: John Hayes, 1796. Evans: 30729 Austin: 1174. Discussion of yellow fever. Mackrill claims to have experience with treating yellow fever in the West Indies; he disputes the theory that the disease is imported. Like Rush he advocates bleeding and purges.

259 **Mitchill, Samuel Latham (1764-1831).** *Address &c.* [New York: 1796]. mp: 47834 Bristol: 9608 Austin: 1317. Proposal to publish collections of medical and other scientific literature.

260 *Observations on Dr. Mackrill's History of the Yellow Fever.* Baltimore: John Hayes, 1796. Evans: 30922 Austin: 1428. A defense of Rush's theories concerning yellow fever, with an attack on Mackrill (see 258).

261 **Otto, John C. (1775-1845).** *An Inaugural Essay on Epilepsy.* Philadelphia: Lang and Ustick, 1796. Evans: 30934 Austin: 1443. Medical dissertation for the University of Pennsylvania; includes description, causes and treatment.

262 [**Pascalis-Ouviere, Felix (c.1750-c.1840)**]. *Medico-Chemical Dissertations on the Causes of the Epidemic Called Yellow Fever and on the Best Antimonial Preparation for the Use of Medicine.* Philadelphia: Snowden and McCorkle, 1796. Evans: 30960 Austin: 1460. Answer to "first prize-question proposed by the Medical Society of Connecticut"; argues against contagion imported from outside the region. Pascalis-Ouviere was a physician from the West Indies who supported the theories of Benjamin Rush.

263 **Plummer, Jonathan, Jr. (1761-1819).** *The Awful Malignant Fever at Newburyport in the Year 1796.* [Newburyport: 1796]. Evans: 31018 Austin: 1548. Account of yellow fever in Newburyport with poems to the dead; Plummer was a well-known Newburyport eccentric and ballad-writer.

264 **Potter, Nathaniel (1770-1843).** *An Essay on the Medicinal Properties and Deleterious Qualities of Arsenic.* Philadelphia: William W. Woodward, 1796. Evans: 31031 Austin: 1558. Medical dissertation for the University of Pennsylvania. Potter was later the first professor of the theory and practice of medicine at the Medical College of Maryland (now the University of Maryland School of Medicine). He also established the non-contagious nature of yellow fever.

265 Ramsay, David (1749-1815). *A Sketch of the Soil, Climate, Weather, and Diseases of South Carolina.* Charleston: W.P. Young, 1796. Evans: 31071 Austin: 1585 Meisel: 3:359. Discussion of preventive medicine covering infants, children, and adults; delivered before the South Carolina Medical Society.

266 Rush, Benjamin (1745-1813). *Medical Inquiries and Observations. Vol. 4.* Philadelphia: Thomas Dobson, 1796. Evans: 31144 Austin: 1660. Collection of Rush's monographs; includes his account of the yellow fever epidemic and a defense of bleeding.

267 [Trusler, John (1735-1820)]. *An Easy Way to Prolong Life by a Little Attention to What We Eat and Drink.* Dover, NH: Samuel Bragg, 1796. Evans: 30381 Austin: 1930. A "chemical analysis" of food and digestion. It includes suggestions for diet.

268 Webster, Noah (1758-1843). *A Collection of Papers on the Subject of Bilious Fevers Prevalent in the United States for a Few Years Past.* New York: Hopkins, Webb, and Co., 1796. Evans: 31593 Austin: 2025. Collection including monographs by Seaman, Elihu Smith, Buel, Taylor and Hansford, Ramsay, Eneas Munson, Mitchill, Reynolds, and Webster himself. Most of the papers, with the exception of that by Munson, argue against the idea that yellow fever is contagious.

269 Wilson, John (d. 1835). *An Inaugural Experimental Dissertation on Digestion.* Philadelphia: Lang and Ustick, 1796. Evans: 31644 Austin: 2075. Medical dissertation for the University of Pennsylvania; with experimental results, including the analysis of vomit.

1797

270 *Address to the Inhabitants of the City and Liberties of Philadelphia.* Philadelphia: 1797. Evans: 31695 Austin: 20. Directions for avoiding yellow fever and recommendations for treatment.

271 Allston, William (d. 1848). *An Inaugural Dissertation on Dropsy, or the Hydropic State of Fever.* Philadelphia: William W. Woodward, 1797. Evans: 31716 Austin: 31. Medical dissertation for the University of Pennsylvania; includes description, causes, and treatment.

272 Andrews, John (1746-1813). *An Address to the Graduates in Medicine, Delivered at Medical Commencement in the University of Pennsylvania, Held May 12, 1797.* Philadelphia: Ormrod and Conrad,

1797. Evans: 31736 Austin: 44. Graduation address; Andrews was Vice Provost of the University.

273 **Bay, William (1773-1865).** *An Inaugural Dissertation on the Operation of Pestilential Fluids upon the Large Intestines, Termed by Nosology, Dysentery.* New York: T. and J. Swords, 1797. Evans: 31780 Austin: 157. Medical dissertation for Columbia; includes description, causes and treatment.

274 **Black, Robert (fl. 1797).** *An Inaugural Dissertation on Fractures.* Philadelphia: Ormrod and Conrad, 1797. Evans: 31832 Austin: 207. Medical dissertation for the University of Pennsylvania.

275 **Boston Dispensary.** *Institution of the Boston Dispensary.* [Boston: 1797]. Evans: 31851 Austin: 235. Description and rules of the Boston Dispensary, a clinic for the poor.

276 **Brown, Samuel (1769-1805).** *An Inaugural Dissertation on the Bilious Malignant Fever.* Boston: Manning and Loring, 1797. Evans: 31881 Austin: 292. Medical dissertation for Harvard; on yellow fever.

277 **Church, John (1774-1806).** *An Inaugural Dissertation on Camphor.* Philadelphia: John Thompson, 1797. Evans: 31935 Austin: 464. Medical dissertation for the University of Pennsylvania; on the nature of camphor and its uses, medical and non-medical.

278 **Cooper, Samuel (1772-1798).** *A Dissertation on the Properties and Effects of the Datura Stramonium or Common Thorn-Apple, and on Its Use in Medicine.* Philadelphia: Samuel H. Smith, 1797. Evans: 31985 Austin: 527. Medical dissertation for the University of Pennsylvania; a description and discussion of uses for the Thorn Apple.

279 **DeWitt, Benjamin (1774-1819).** *A Chemico-Medical Essay to Explain the Operation of Oxigene or the Base of Vital Air on the Human Body.* Philadelphia: William W. Woodward, 1797. Evans: 32036 Austin: 665. Medical dissertation for the University of Pennsylvania; on oxygen and respiration.

280 **Fisher, James (1774-1802).** *An Inaugural Dissertation on That Grade of the Intestinal State of Fever Known by the Name of Dysentery.* Philadelphia: Ormrod and Conrad, 1797. Evans: 32131 Austin: 774. Medical dissertation for the University of Pennsylvania; includes description, causes, and treatment.

281 **Fleet, John Jr. (1766-1813).** *A Discourse Relative to the Subject of Animation, Delivered before the Humane Society of the Commonwealth of Massachusetts at Their Semiannual Meeting, June 13th.* Boston: John and Thomas Fleet, 1797. Evans: 32134 Austin:

777. Discussion of the origin of life in humans, relative to "restoring life" after accidents.

282 [Folwell, Richard (c.1768-1814)]. *Short History of the Yellow Fever That Broke Out in the City of Philadelphia in July, 1797.* Philadelphia: Richard Folwell, 1797. Evans: 32138 Austin: 781. Description of epidemic of 1797, including list of the dead. According to Austin, part of the documents Folwell intended to include were suppressed "because of pending legal suits."

283 Hosack, Alexander, Jr. (d. 1834). *History of the Yellow Fever As It Appeared in the City of New York in 1795.* Philadelphia: Thomas Dobson, 1797. Evans: 32282 Austin: 952. Popular version of Hosack's dissertation (see 284), including symptoms, causes, and cures (no attempt to discuss origin).

284 Hosack, Alexander, Jr. (d. 1834). *An Inaugural Essay on the Yellow Fever As It Appeared in This City in 1795.* New York: T. and J. Swords, 1797. Evans: 32283 Austin: 953. Medical dissertation for Columbia; includes symptoms, causes, and cure.

285 Huger, Francis Kinlock (1773-1855). *An Inaugural Dissertation on Gangrene and Mortification.* Philadelphia: Stephen C. Ustick, 1797. Evans: 32289 Austin: 989. Medical dissertation for the University of Pennsylvania; includes description and treatment.

286 Johnson, Joseph (1776-1862). *An Experimental Inquiry into the Properties of Carbonic Acid Gas or Fixed Air, Its Mode of Operation, Use in Diseases, and Most Effectual Method of Relieving Animals Affected by It.* Philadelphia: Stephen C. Ustick, 1797. Evans: 32319 Austin: 1071. Medical dissertation for the University of Pennsylvania; on carbon dioxide.

287 Jones, Samuel (1772-1797). *An Inaugural Dissertation on Hydrocele.* Philadelphia: Samuel C. Ustick, 1797. Evans: 32324 Austin: 1088. Medical dissertation for the University of Pennsylvania; on hernia.

288 Laws, John. *An Inaugural Dissertation on the Rationale of the Operation of Opium on the Animal Economy, with Observations on Its Use in Disease.* Wilmington: W.C. Smyth, 1797. Evans: 32364 Austin: 1134. Medical dissertation for the University of Pennsylvania.

289 Mackenzie, Colin (1775?-1827). *An Inaugural Dissertation on the Dysentery.* Philadelphia: Ormrod and Conrad, 1797. Evans: 32407 Austin: 1172. Medical dissertation for the University of Pennsylvania; includes description, causes, and treatment.

290 [Mitchill, Samuel Latham (1764-1831)]. *The Case of the Manufacturers of Soap and Candles in the City of New York, Stated and Examined.* New York: John Buel, 1797. Evans: 32564 Austin: 90. Description of the New York laws against unhealthy manufacturing and a defense of the soap and tallow manufacturers against the charge that their processes cause yellow fever. Mitchill argues against Rush's "bad air" hypothesis regarding the origin of yellow fever. An appendix includes citations of various authorities.

291 Mitchill, Samuel Latham (1764-1831). *The Present State of Medical Learning in the City of New York.* New York: T. and J. Swords, 1797. Evans: 32488 Austin: 1322. Discussion of medical education offered in New York through Columbia and New York Hospital; includes history of Columbia medical school, current curriculum, and description of hospital facilities.

292 New York Dispensary. *Charter and Ordinances of the New York Dispensary.* New York: Hopkins, Webb, and Co., 1797. Evans: 32566 Austin: 1383. Incorporation act of 1795; includes accounting and some statistics on patients.

293 New York Hospital. *No. 1.* [New York: 1797] mp: 48198 Bristol: 10031. Description of charity hospital, with accounts and description of diseases treated.

294 North, Edward Washington (d. 1843). *An Inaugural Dissertation on the Rheumatic State of Fever.* Philadelphia: William W. Woodward, 1797. Evans: 32592 Austin: 1424. Medical dissertation for the University of Pennsylvania; includes description, causes, and treatment.

295 Pennsylvania Hospital. *The Committee Appointed to Prepare an Account of the Monies Received from the Legislature Report. . .* [Philadelphia: 1797]. Bristol: 10059 Not included in Readex. Description of expenses.

296 Pennsylvania Hospital. *The Committee Appointed to Prepare an Account of the Monies Received from the Legislature of Pennsylvania towards Erecting Additional Buildings to the Pennsylvania Hospital and Finishing the Same, and of the Expenditures of Said Buildings, Report.* Philadelphia: John Fenno, 1797. Evans: 32680 Austin: 1526. Description of expenses in constructing new buildings for the Pennsylvania Hospital.

297 Pennsylvania Hospital. *To the Senate and House of Representatives of the Commonwealth of Pennsylvania.* [Philadelphia: 1797]. mp: 48219 Bristol:1006. History of Pennsylvania Hospital with request for additional funds.

298   Perkins, Elisha (1741-1799). *Evidences of the Efficacy of Doctor Perkins's Patent Metallic Instruments.* New Haven: B. Green, [1797]. Evans: 32668 Austin: 1492. Testimonials regarding the success of Perkins' instruments. The instruments were pieces of metal which were stroked above the "afflicted area"; Perkins claimed to apply the principles of Galvani to medical practice. The instruments were quite popular in America and Britain in the late eighteenth century.

299   Philadelphia Dispensary. *To the Attending Physician. . . .* [Philadelphia: 1797]. mp: 48338 Bristol: 10069 Austin: 1514. Referral form for charity patient; includes dispensary rules.

300   Spalding, Lyman (1775-1821). *An Inaugural Dissertation on the Production of Animal Heat.* Walpole: David Carlisle, Jr., 1797. Evans: 32864 Austin: 1804. Medical dissertation for Harvard.

301   Stewart, Jesse. *Genuine French Creek Seneca Oil.* [Springfield, NJ?: 1797?] mp: 48258 Bristol: 10114 Austin: 816. Advertisement for patent medicine with description of its uses; includes testimonials.

302   Stock, John Edmonds. *An Inaugural Dissertation on the Effects of Cold upon the Human Body.* Philadelphia: Joseph Gales, 1797. Evans: 32883 Austin: 1829. Medical dissertation for the University of Pennsylvania; Stock was from Britain, a "member of the medical and natural history societies of Edinburgh."

303   Taplin, William (d. 1807). *A Compendium of Practical and Experimental Farriery.* Wilmington: Bonsal and Niles, 1797. Evans: 32906 Austin: 1856. Veterinary guide for horses.

304   Vaughan, John (1775-1807). *Observations on Animal Electricity in Explanation of the Metallic Operation of Dr. Perkins.* Wilmington: W.C. Smyth, 1797. Evans: 33112 Austin: 1975. Testimonial for Perkins' instruments (see 298).

305   Walker, James. *An Inquiry into the Causes of Sterility in Both Sexes with Its Method of Cure.* Philadelphia: E. Oswald, 1797. Evans: 33143 Austin: 1986. Medical dissertation for the University of Pennsylvania.

306   Wilson, Goodridge. *An Inaugural Dissertation on Absorption.* Philadelphia: E. Oswald, 1797. Evans: 33233 Austin: 2072. Medical dissertation for the University of Pennsylvania; on the absorption capacity of animals.

1798

307   Academy of Medicine of Philadelphia. *Proofs of the Origin of the Yellow Fever in Philadelphia and Kensington in the Year 1797, from Domestic Exhalation.* Philadelphia: Thomas and Samuel

Bradford, 1798. Evans: 34352 Austin: 7. Supports Rush's theory that cause of yellow fever is "putrid exhalations"; the monograph is signed by Rush and twelve others, including Otto, Caldwell, John Redman Coxe, Sayre, Cooper, and Pascalis-Ouviere. The Academy was founded by Rush to combat the Philadelphia College of Physicians.

308 Archer, John Jr. (1777-1830). *An Inaugural Dissertation on Cynanche Trachealis, Commonly Called Croup or Hives.* Philadelphia: Way and Groff, 1798. Evans: 33307 Austin: 50. Medical dissertation for University of Pennsylvania; includes description, causes, and treatment.

309 Aspinwall, George (1774-1804). *A Dissertation on the Cynanche Maligna.* Dedham: Mann and Adams, 1798. Evans: 33318 Austin: 88. Discussion of history, causes and treatment; rejects bleeding and "antiphlogiston" treatments, suggests cause is "contaminated air."

310 Barton, Benjamin Smith (1766-1815). *Collections for an Essay towards a Materia Medica of the United States.* Philadelphia: Way and Groff, 1798. Evans: 33377 Austin: 137 Hazen: 1513 Meisel: 3:360. Speech delivered to Philadelphia Medical Society; suggests establishing the medical properties of American "vegetables and plants."

311 Browne, Joseph (fl. 1798). *A Treatise on the Yellow Fever, Shewing Its Origin, Cure, and Prevention.* New York: Argus, 1798. Evans: 33465 Austin: 299. Discussion of yellow fever; Browne favors an environmental cause, rather than an imported one.

312 Burrell, William. *Medical Advice, Chiefly for the Consideration of Seamen.* New York: R. Wilson, 1798. mp: 48383 Bristol:10250 Austin: 368. Collection of essays on prevention and cure of disease and injury on ship, including gun-shot wounds, fractures, dislocations, and venereal disease.

313 Claiborne, John (d. 1808). *An Inaugural Essay on Scurvy.* Philadelphia: Stephen C. Ustick, 1798. Evans: 33518 Austin: 467. Medical dissertation for the University of Pennsylvania. Includes description, causes, and treatment.

314 Cocke, William. *An Inaugural Dissertation on Tetanus.* Philadelphia: R. Atiken, 1798. Evans: 33530 Austin: 488. Medical dissertation for the University of Pennsylvania. Includes description, causes, and treatment.

315 Currie, William (c.1754-1828). *Memoirs of the Yellow Fever, Which Prevailed in Philadelphia and Other Parts of the United States of America in the Summer and Autumn of the Present Year, 1798.* Philadelphia: John Bioren, 1798. Evans: 33589 Austin: 602. History of the 1798 epidemic, including weather, accounts of the dead, and results of dissections. Written in running journal form. Currie still favors foreign origin of disease.

316 Currie, William (c.1754-1828). *Observations on the Causes and Cure of Remitting or Bilious Fevers.* Philadelphia: William T. Palmer, 1798. Evans: 33590 Austin: 603. Discussion of "situations, climates, and seasons" when "remitting or bilious fevers" are most prevalent. Currie closes with an appendix on yellow fever in which he distinguishes it from bilious fever.

317 Currie, William (1754-1828). *Of the Cholera.* [Philadelphia: William T. Palmer, 1798]. Evans: 33591 Austin: 604. Discussion of symptoms, causes, and treatment of cholera.

318 Davidge, John Beale (1768-1829). *A Treatise on the Autumnal Endemial Epidemic of Tropical Climates, Vulgarly Called the Yellow Fever* Baltimore: W. Pechin, 1798. Evans: 33603 Austin: 626. Discussion of Baltimore yellow fever epidemic; Davidge supported the idea of an environmental cause. According to the DAMB, Davidge "made a major contribution to public discussion (lay and professional) on causes of yellow fever."

319 Dick, Elisha Cullen (1762-1825). *Doctor Dick's Instructions for the Nursing and Management of Lying-In Women.* Alexandria: Thomas and Westcott, 1798. Evans: 33635 Austin: 668. Requirements for delivery room and treatment of women before, during, and after delivery; includes directions for treatment of newborns. Dick is best known as one of Washington's doctors during his final illness.

320 Disborough, Henry. *An Inaugural Dissertation on Cholera Infantum.* Philadelphia: Budd and Bartram, 1798. Evans: 33641 Austin: 673. Medical dissertation for University of Pennsylvania; includes description, causes and treatment.

321 Hahn, John. *Observations and Experiments on the Use of Enemata and the External Application of Medicines to the Human Body.* Philadelphia: Stephen C. Ustick, 1798. Evans: 33831 Austin: 849. Medical dissertation for the University of Pennsylvania; on enemas and medicines absorbed through the skin.

322 Hoffman, Christian. *Longevity, Being an Account of Various Persons Who Have Lived to an Extraordinary Age with Several Curious Particulars Reflecting Their Lives.* New York: Jacob S. Mott, 1798. Evans: 33887 Austin: 920. List of various people reported to have lived over 100 years, with the source of the report; includes scriptural accounts in an appendix.

323 Holt, Charles (1772-1852). *A Short Account of the Yellow Fever As It Appeared in New London in August, September, and October, 1798.* New London: Charles Holt, 1798. Evans: 33890 Austin: 927. History of 1798 New London yellow fever epidemic; includes list of the dead.

324 Horsfield, Thomas (1773-1859). *An Experimental Dissertation on the Rhus Vernix, Rhus Radicans and Rhus Glabrum, Commonly Known in Pennsylvania by the Names of Poison Ash, Poison Vine, and Common Sumach.* Philadelphia: Charles Cist, 1798. Evans: 33905 Austin: 951. Discussion of three native plants. According to the DAB, it is "remarkable for its painstaking description of the toxic symptoms of poisoning produced by sumac and poison ivy, and for the record of well-conceived experiments, carried out upon himself and upon animals. . . .It ranks as a pioneer contribution in the history of experimental pharmacology in America."

325 Hubard, James Thruston. *An Inaugural Dissertation on Puerpal Fever.* Philadelphia: John Ormrod, 1798. Evans: 33908 Austin: 987. Medical dissertation for the University of Pennsylvania; includes description, causes, and treatment.

326 Johnson, Thomas (fl. 1773-1798). *Every Man His Own Doctor or the Poor Man's Physician.* Salisbury: 1798. mp: 48490 Bristol: 10369 Austin: 1076. Collection of home remedies; the title and part of the preface are taken from John Tennent (see 24), but the rest of the collection is new.

327 Lent, Adolph C. *An Inaugural Dissertation, Shewing in What Manner Pestilential Vapours Acquire Their Acid Quality, and How This Is Neutralized and Destroyed by Alkalies.* New York: T. and J. Swords, 1798. Evans: 33992 Austin: 1141. Medical dissertation for Columbia; on "pestilential vapours," i.e., effluvia from animal and vegetable decomposition.

328 *Observations on the Influence of the Moon on Climate and the Animal Economy with a Proper Method of Treating Diseases When under the Power of That Luminary.* Philadelphia: Richard Folwell, 1798. Evans: 34264 Austin: 1429. Discussion of the moon's influence on disease; argues that the moon affects climate and health.

329 Pascalis-Ouviere, Felix (c.1750-c.1840). *An Account of the Contagious Epidemic Yellow Fever Which Prevailed in Philadelphia in the Summer and Autumn of 1797.* Philadelphia: Snowden and McCorkle, 1798. Evans: 34311 Austin: 1458. History of yellow fever prior to Philadelphia outbreak; description of disease and treatment, with case studies.

330 Pennsylvania. *Compilation of the Health Laws of the State of Pennsylvania.* Philadelphia: Zachariah Poulson, 1798. Evans: 34324 Austin: 1486. All the state laws pertaining to health.

331 Pennsylvania. *Letter from the Secretary of the Commonwealth of Pennsylvania by Direction of the Governor, Relative to the Late Malignant Fever, and Report of the Board of Managers of the Marine and City Hospitals in Reply.* Philadelphia: Thomas and Samuel Bradford, 1798. Evans: 34332 Austin: 1489. Letter to health office with questions about yellow fever epidemic preliminary to report on "mortality and expenditure." Includes reply with some preliminary answers.

332 Philadelphia College of Physicians. *Facts and Observations Relative to the Nature and Origin of the Pestilential Fever Which Prevailed in This City in 1793, 1797, and 1798.* Philadelphia: Thomas Dobson, 1798. Evans: 34355 Austin: 497. History of yellow fever, maintaining the disease is of foreign origin and contagious. Austin calls it a "description of the college's actions during the yellow fever epidemic."

333 Philadelphia College of Physicians. *Proceedings of the College of Physicians of Philadelphia Relative to the Prevention of the Introduction and Spreading of Contagious Diseases.* Philadelphia: Thomas Dobson, 1798. Evans: 34356 Austin: 498. Minutes, correspondence, and quarantine recommendations; supporting college's claim that yellow fever is imported from West Indies and contagious.

334 *Physical Enquiry into the Origin and Causes of the Pestilential Fever.* New York: J. Tiebout, 1798. Evans: 34377 Austin: 1544. Argument in favor of putrefying animal and vegetable material as causes of yellow fever.

335 Prioleau, Philip Gendron (d. 1814). *An Inaugural Dissertation on the Use of the Nitric and Oxigenated Muriatic Acids in Some Diseases.* Philadelphia: John Bioren, 1798. Evans: 34418 Austin: 1567. Medical dissertation for the University of Pennsylvania.

336 Rush, Benjamin (1745-1813). *Medical Inquiries and Observations, Containing an Account of the Yellow Fever As It Appeared in Philadelphia in 1797, and Observations upon the Nature and Cure of the Gout and Hydrophobia. Vol. V.* Philadelphia: Budd and Bartram, 1798. Evans: 34496 Austin: 1660. Collection of medical monographs, largely concerned with yellow fever.

337 Rush, Benjamin (1745-1813). *A Syllabus of a Course of Lectures on the Institutes and Practice of Medicine.* Philadelphia: Thomas and Samuel Bradford, 1798. Evans: 34497 Austin: 1689. Outline of lecture topics. Rush had succeeded John Morgan as professor of the theory and practice of medicine at the University of Pennsylvania in 1789; he became professor of the institutes of medicine and clinical practice as well in 1792.

338 Salem (MA) Hospital. *The Following Regulations Are to Be Observed at Salem Hospital.* [Salem: 1798]. mp: 48607 Bristol: 10510. Hospital rules.

339 Smith, James (1738-1812). *A Concise, Economical Plan of the Family Medical Institution for Administering Advice and Medicines to Families and Individuals Possessing Small Fortunes and Moderate Incomes.* New York: T. Kirk [1798?]. mp: 48617 Bristol: 10523 Austin: 1762. Proposal for establishing a "medical institution" under Smith's direction.

340 Smith, James (1738-1812). *Yellow Fever.* [New York: 1798] mp: 48618 Bristol: 10524 Austin: 1763. Chiefly discussion of causes (environmental); recommended cure is attendance of good physician.

341 [Stearns, Samuel (1748-1819)]. *Sir. . . .* [Providence: 1798]. Evans: 34602 Austin: 1819. Open letter to members of Rhode Island General Assembly proposing establishment of a lottery to subsidize the production of free medical texts reflecting the latest discoveries.

342 Stuart, James. *A Dissertation on the Salutary Effects of Mercury in Malignant Fevers.* Philadelphia: Thomas and Samuel Bradford, 1798. Evans: 34619 Austin: 1835. Medical dissertation for University of Pennsylvania; on medicinal use of mercury.

343 Triplett, Thomas. *An Inaugural Dissertation on Apoplexy.* Philadelphia: Way and Groff, 1798. Evans: 34680 Austin: 1925. Medical dissertation for the University of Pennsylvania; includes description, causes, and treatment.

344 Webb, William (fl. 1798). *An Inaugural Dissertation on the Colic.* Philadelphia: John Ormrod, 1798. Evans: 34974 Austin: 2022. Medical dissertation for the University of Pennsylvania; includes description, causes and treatment.

345 Winston, Isaac. *An Inaugural Dissertation on the Polygala Senega, Commonly Called Seneca Snake Root.* Philadelphia: Way and Groff, 1798. Evans: 35042 Austin: 2080. Medical dissertation for the University of Pennsylvania; experimental analysis of snake root, citing Tennent (see 27).

346 Yeatman, Charleton. *The Mariner's Guide, or Plain Instructions to Them How to Treat Every Disease Which They Are Liable to, with Their General Symptoms to Know One Disorder from Another, to Accompany Medicine Chests, Put Up Agreeably to an Act of Congress.*

Philadelphia: John Bioren, 1798. Evans: 35066 Not included in
Readex Austin: 2101. Apparently a self-treatment guide; only recorded
copy is lost.

1799

347 Adams, Daniel (1773-1864). *An Inaugural Dissertation on the
Principle of Animation.* Hanover, NH: Moses Davis, 1799. Evans:
35074 Austin: 18. Medical dissertation for "Bachelor of Medicine" at
Dartmouth.

348 Bayley, Richard (1745-1801). *Letters from the Health Office,
Submitted to the Common Council of the City of New York.* [New
York]: John Furman, [1799]. Evans: 35161 Austin: 162. Letters
concerning the epidemiology of yellow fever. Bayley was physician to
the Port of New York; he helped formulate the United States and New
York quarantine laws.

349 Bellinger, John Skottowe. *An Inaugural Dissertation on
Chronic Pneumony [sic] or Pulmonary Consumption.* Philadelphia:
Way and Groff, 1799. Evans: 35177 Austin: 199. Medical
dissertation for the University of Pennsylvania; includes description,
causes, and treatment.

350 Boston Board of Health. *Board of Health to Their Constituents.*
[Boston: 1799]. Evans: 35220 Austin: 223. Health regulations
regarding epidemics; Paul Revere was Chairman of the Board.

351 Brailsford, Edward (d. 1856). *An Experimental Dissertation
on the Chemical and Medical Properties of the Nicotiana Tabacum of
Linnaeus, Commonly Known by the Name of Tobacco.* Philadelphia:
John Ormrod, 1799. Evans: 35230 Austin: 267. Medical dissertation
for the University of Pennsylvania; discusses physical and "moral"
effects of tobacco use, including a chemical analysis. Includes a
drawing of a tobacco plant.

352 Brevitt, Joseph (1769-1839). *The History of Anatomy from
Hippocrates, Who Lived Four Hundred Years Before Christ.*
Baltimore: Samuel Sower, 1799. Evans: 35232 Austin: 272.
Biographies of famous anatomists.

353 Caldwell, Charles (1772-1853). *An Eulogium to the Memory
of Dr. Samuel Cooper.* Philadelphia: Henry Tuckniss, 1799. Evans:
35262 Austin: 387. Delivered to the Philadelphia Medical Society;
includes biography of Cooper.

354 Caldwell, Charles (1772-1853). *A Semi-Annual Oration on
the Origin of Pestilential Diseases.* Philadelphia: Thomas and Samuel
Bradford, 1799. Evans: 35263 Austin: 395. Delivered to Academy of

Medicine of Philadelphia; suggests that yellow fever is caused by foul air from "putrid ballast" or "damaged cargoes" of ships.

355 Chalwill, William G. *A Dissertation on the Sources of Malignant Bilious or Yellow Fever, and the Means of Preventing It.* Philadelphia: Way and Groff, 1799. Evans: 35291 Austin: 436. Medical dissertation for the University of Pennsylvania; Chalwill favors environmental causes for yellow fever.

356 Church, James. *The Efficacy of Dr. Church's Cough Drops, Demonstrated in the Cure of Coughs, Colds, Asthma, and Consumptions.* New York: G. and R. Waite, 1799. Evans: 35305 Austin: 462. Advertisement for patent medicine; includes testimonial affidavits and advertisements for other Church's medicines.

357   Condie, Thomas (c.1775-1814) and Richard Folwell (c.1768-1814). *History of the Pestilence Commonly Called Yellow Fever, Which Almost Desolated Philadelphia in the Months of August, September, and October, 1798.* Philadelphia: R. Folwell, [1799]. Evans: 35335 Austin: 516. Description of epidemic of 1798; includes information about epidemics in other cities. Appendixes cover Academy of Medicine letter, Board of Health letters, and names of dead.

358 Foushee, John H. *An Inaugural Essay on Strictures in the Urethra.* Philadelphia: John Ormrod, 1799. Evans: 35494 Austin: 787. Medical dissertation for the University of Pennsylvania.

359 Giraud, Jean Jacques (1759-1839). *Dr. Giraud's Specific and Universal Salt for the Venereal Disease and All the Venereal Affections Which Are the Result of It.* [Philadelphia: 1799]. mp: 48861 Bristol: 10792 Austin: 822. Description of patent medicine.

360 Hardie, James (1758-1826). *An Account of the Malignant Fever Lately Prevalent in the City of New York.* New York: Hurtin and McFarlane, 1799. Evans: 35586 Austin: 877. History of New York yellow fever epidemic of 1798, including a list of the dead.

361 King, Robert J. *An Inaugural Dissertation on Blisters.* Philadelphia: Way and Groff, 1799. Evans: 35687 Austin: 1095. Medical dissertation for the University of Pennsylvania; on the use of cantharides to blister the skin as treatment.

362 Lining, John (1708-1760). *A Description of the American Yellow Fever Which Prevailed at Charleston, in South Carolina, in the Year 1748.* Philadelphia: Thomas Dobson, 1799. Evans: 35733 Austin: 1155. Originally written in 1753 as a letter to Dr. Robert Whytt of Edinburgh, where it was published. It is "the earliest American account of the symptoms and pathology of yellow fever" (DAB).

363 May, Arthur (d. 1812). *An Inaugural Dissertation on Sympathy.* Philadelphia: Way and Groff, 1799. Evans: 35807 Austin: 1241. Medical dissertation for University of Pennsylvania; on the effect of one disease in curing another.

364 New Hampshire Medical Society, Eastern District. *Laws of the Eastern District of the New Hampshire Medical Society.* Exeter: H. Ranlet, 1799. Evans: 35891 Austin: 1370. Rules of the Society.

365 New York (City) Committee on Health. *Record of Death.* New York: John Hill, [1799]. Evans: 35942 Austin: 1388. List of names, addresses, occupations and death dates of yellow fever victims for the 1799 epidemic.

366 New York (State) Medical Society. *Report of the Committee Appointed by the Medical Society of the State of New York to Enquire into the Symptoms, Origin, Cause and Prevention of the Pestilential Disease That Prevailed in New York during the Summer and Autumn of the Year 1798.* New York: 1799. Evans: 35933 Austin: 1279. Report of committee (including S.L. Mitchill, James Tillary, and John R.B. Rodgers) on yellow fever epidemic. The committee favored an environmental cause.

367 Norcom, James (1778-1850). *An Inaugural Thesis on Jaundice, Containing Observations on the Liver and Some of Its Diseases.* Philadelphia: James Carey, 1799. Evans: 35977 Austin: 1423. Medical dissertation for the University of Pennsylvania; includes description, causes, and treatment.

368 Pennsylvania Hospital. *To the Senate and House of Representatives of the Commonwealth of Pennsylvania.* Philadelphia: Zachariah Poulson, Jr., 1799. Evans: 36104 Austin: 1530. Account of receipts and expenditures, with plea that new construction continue.

369 Philadelphia Academy of Medicine. *Constitution of the Academy of Medicine of Philadelphia.* [Philadelphia: 1799]. Evans: 36093 Austin: 5. Includes purpose and aims of Academy and reasons for incorporation (chiefly research-oriented).

370 Philadelphia Academy of Medicine. *Laws of the Academy of Medicine of Philadelphia.* [Philadelphia: 1799]. Evans: 36094 Austin: 6. By-laws of the Academy.

371 Rush, Benjamin (1745-1813). *Observations upon the Origin of the Malignant Bilious or Yellow Fever in Philadelphia and upon the Means of Preventing It.* Philadelphia: Budd and Bartram, 1799. Evans: 36253 Austin: 1675. Reiteration of Rush's belief that yellow fever was caused by a "miasma" of "domestic origin." However, according to

Winslow, the pamphlet also represents a retreat from Rush's earlier contention that the disease was contagious.

372   Rush, Benjamin (1745-1813). *A Second Address to the Citizens of Philadelphia Containing New Proofs of the Domestic Origin of the Malignant Bilious or Yellow Fever.* Philadelphia: Budd and Bartram, 1799. Evans: 36254 Austin: 1679. More opposition to an imported origin for the disease; the pamphlet also shows Rush's opposition to maritime quarantine.

373   Say, Benjamin (1755-1813). *An Annual Oration Pronounced before the Humane Society of Philadelphia on the Objects and Benefits of Said Institution.* Philadelphia: Whitehall, 1799. Evans: 36278 Austin: 1708. History of the society, including international origins and the development of the Philadelphia society itself. The pamphlet includes the principles and practice of artificial respiration.

374 *Select Pamphlets Respecting the Yellow Fever.* Philadelphia: Mathew Carey, 1799. Evans: 36287 Austin: 1728. Contains monographs by Mathew Carey, Jean Deveze, William Currie, Richard Folwell, and Thomas Condie.

375   Society of the Lying-In Hospital of the City of New York. *An Act to Incorporate the Society of the Lying-In Hospital in the City of New York.* Brooklyn: T. Kirk, [1799]. Evans: 35953 Austin: 1786. Includes hospital rules and by-laws.

376   Society of the Lying-In Hospital of the City of New York. *Constitution of the New York Lying-In Hospital.* [New York]: John Furman, 1799. Evans: 35954 Austin: 1787. Description of organization.

377   Watts, Washington. *An Inquiry into the Causes and Nature of the Yellow Fever.* Philadelphia: John Ormrod, 1799. Evans: 36678 Austin: 2018. Medical dissertation for the University of Pennsylvania; favors environmental cause.

378   Webster, Noah (1758-1843). *Brief History of Epidemic and Pestilential Diseases. Vols. 1 and 2.* Hartford: Hudson and Goodwin, 1799. Evans: 36687 Austin: 2023. According to Winslow, "Its two volumes represent. . . so far as I am aware, the best general summary of epidemiological opinion at the beginning of the nineteenth century. . . and few works surpass it as a compendium of earlier speculations in the field."

379   Wheaton, Jesse. *Wheaton's Genuine Jaundice Bitters, Secured to Him by Letters Patent from the President of the United States and Prepared and Sold by Him in Dedham, MA.* [Dedham: 1799]. mp:

48991 Bristol: 10948 Austin: 2035. Advertisement for patent medicine; includes testimonials.

Technical Science Titles

1705

380 [Bradford, William (1658-1752)]. *The Secretary's Guide or Young Man's Companion. In Four Parts.* 4th ed. New York: William Bradford, 1705. Evans: 2997 Karpinsky: p. 36, 39 Guerra: a-155. Textbook including a section on "Arithmetick Made Easie." Karpinsky calls it the first serious discussion of arithmetic published in America; it is based almost entirely on two English textbooks by William Mather and William Leybourn.

1717

381 Southack, Cyprian (1662-1745). *A New Chart of North America.* Boston: Francis Dewing, 1717. Not included in Readex Wheat/Brun: 44. According to Wheat and Brun, the first general map of eastern North American continent printed in America. It was published to demonstrate the French threat to North America by showing the location of French forts.

1720

382 Southack, Cyprian (1662-1745). *The Harbours and Islands of Canso, Part of the Boundaries of Nova Scotia.* Boston: Francis Dewing, 1720. Not included in Readex Wheat/Brun: 68. Southack was one of the commissioners sent to adjust the Nova Scotia boundaries in 1718; he was a member of the Council in Nova Scotia in 1720.

1722

383 Bonner, John (c. 1643-1725/6). *The Town of Boston in New England.* Boston: Francis Dewing, 1722. Evans: 2318 Wheat/Brun: 224. Reprinted at least nine times over 47 years; remained in use until Osgood Carleton's map of 1796 (see 627). Wheat and Brun call it "the earliest and most important engraved plan of Boston."

1729

384 [Greenwood, Isaac (1702-1745)]. *Arithmetick [sic] Vulgar and Decimal: with the Application Thereof to a Variety of Cases in Trade and Commerce.* Boston: S. Kneeland, 1729. Evans: 3170 Karpinsky: p. 43. First mathematics textbook written in English by a native American. It was published anonymously, but Greenwood was identified in later newspaper advertisements.

1731

385 Lyne, James. *A Plan of the City of New York from an Actual Survey.* [New York]: William Bradford, [1731]. Evans: 3438 Wheat/Brun: 378. Dedicated to John Montgomery; shows New York as laid out by the Montgomery charter of 1730-31. Wheat and Brun call it "probably the first plan of the City published in New York and one of the first examples of copper engraving made there."

1732

386 *The Complete Mariner.* Williamsburg: E.L. James Hubbard, 1732.
mp: 39987 Bristol: 867. Only the title page survives; it is subtitled "A
treatise of navigation trigonometrically by logaritmetical numbers and
the geometrical construction by scale and compass. Also the
orthographic projection of the sphere astronomical."

1735

387 Colden, Cadwallader (1688-1766). *A Map of the Country of
the Five Nations and a Paper Relating to an Act of Assembly of the
Province of New York, for Encouragement of the Indian Trade, & c.
and for Prohibiting the Selling of Indian Goods to the French, viz. of
Canada.* New York: William Bradford, 1735. Evans: 3921. Map of
Northern New York and an essay and correspondence concerning the
fur trade. It was Colden's first important work, written in 1727, but
enlarged later; it is based entirely on French sources (although Colden
was closely associated with the New York Indian tribes) and was
widely read.

388 [Greenwood, Isaac (1702-1745)]. *A Course of Mathematical
Lectures and Experiments.* [Boston: c.1735]. mp: 39937 Bristol: 809.
Outline of lectures as prospectus for subscribers; the subject matter is
Newtonian theory.

389 *A New Map of the Harbour of New York by a Late Survey.* New
York: William Bradford, [1735.] Evans: 3922 Wheat/Brun: 379.
Similar to a map published by Henry Popple in London; local place
names have been added.

1744

390 Franklin, Benjamin (1706-1790). *An Account of the New
Invented Pennsylvanian Fireplace: Wherein Their Construction and
Manner of Operation Is Particularly Explained; Their Advantages above
Every Other Method of Warming Rooms Demonstrated and All
Objections That Have Been Raised against the Use of Them, Answered
and Obviated.* Philadelphia: Benjamin Franklin, 1744. Evans: 5395
Miller: 349. A description of the Franklin stove with parts list,
directions for assembly and operation, and a diagram. Franklin also
lists its advantages over other designs.

1745

391 [Johnston, Thomas (c. 1708-1767)]. *A Plan of Cape Breton
and Fort Louisbourgh, &c.* n.p.:1745. Not included in Readex
Wheat/Brun: 71. According to Wheat and Brun this map "has the
appearance of being an early engraving by Thomas Johnston," a Boston
painter and engraver.

1746

392   Johnston, Thomas (1708-1767). *A Chart of Canada River from the Islands of Anticosty As Far up As Quebeck [sic], the Islands, Rocks, Shoals, and Soundings As They Appear at Low Water.* Boston: Thomas Johnston, 1746. Not included in Readex Wheat/Brun: 76. Includes many place names along the Canada or St. Lawrence river.

393   Pelham, Paul (c.1695-1751). *A Plan of the City and Fortress of Louisbourg with a Small Plan of the Harbour.* Boston: 1746. Not included in Readex Wheat/Brun:75. Pelham was a painter and engraver in Boston best known for his portraits of New England clergy.

1747

394   Burnham, Jonathan. *A Small Tract of Arithmetick [sic] for the Use of Farmers and Country People.* New London: T. Green, 1747. Evans: 6106 Karpinsky: p. 59. Basic arithmetic text; examples deal with practical problems such as land measurement and debt payment.

1748

395   Eliot, Jared (1685-1763). *An Essay upon Field Husbandry in New England As It Is or May Be Ordered.* New London: T. Green, 1748. Evans: 6132. First of Eliot's popular series of monographs on agriculture; includes discussions of cultivation, land preparation, wool and linen trade, iron manufacture and hemp. Eliot asks for advice from other farmers and proposes publishing their results.

1749

396   Eliot, Jared (1685-1763). *Continuation of the Essay upon Field Husbandry As It Is or May Be Ordered in New England.* New London: T. Green, 1749. Evans: 6313. Second of Eliot's six essays; includes discussion of historical agriculture with biblical references. It also includes much agricultural advice and Franklin's endorsement of Eliot's work.

397   Evans, Lewis (1700-1756). *A Map of Pennsylvania, New Jersey, New York, and the Three Delaware Counties.* Philadelphia: 1749. Evans: 6316 Wheat/Brun: 295. Includes lengthy notes engraved on the map giving information about climate, navigation, and other topics. It includes a statement that American storms "begin leeward," which is considered the "first recorded statement" (Wheat/Brun) that American storms travel in an easterly direction. It was also notable for the detail of the roads around Lancaster, York, and Carlisle, which were the primary route for the migrations to the Carolinas and Tennessee (DAB).

1751

398   Eliot, Jared (1685-1763). *A Continuation of the Essay on Field-Husbandry As It Is or May Be Ordered in New England; with an Appendix by Ebeneezer Silliman.* New London: T. Green, 1751.

Evans: 6666. Part three of Eliot's discussion; includes discussion of flax cultivation and the cultivation of various grains. Silliman's appendix is on hay-making.

1752

399 Heap, George and Nicholas Scull (1687-1761). *A Map of Philadelphia and Parts Adjacent with a Perspective View of the State-House.* [Philadelphia: 1752]. Evans: 7209 Wheat/Brun: 454. Wheat and Brun describe this map as "the most popular ever issued." Many reproductions followed.

1753

400 Grew, Theophilus (d.1759). *The Description and Use of the Globes, Celestial and Terrestrial; with Variety of Examples for the Learner's Exercise: Intended for the Use of Such Persons As Would Attain to the Knowledge of Those Instruments; but Chiefly Designed for the Instruction of the Young Gentlemen at the Academy in Philadelphia, to Which Is Added Rules for Working All the Cases in Plain and Spherical Triangles Without a Scheme.* Germantown: Christopher Sower, 1753. Evans: 7012 Karpinsky: p. 63. According to Karpinsky, the "first trigonometric treatise of the Americas." Grew was the first professor of mathematics at the College of Philadelphia (University of Pennsylvania).

401 Johnston, Thomas (c. 1708-1767). *A True Coppy [sic] from an Ancient Plan of E. Hutchinson's Esqr. and from Jose. Heath in 1719 and Phins. Jones's Survey in 1731 and from John North's Late Survey in 1752.* Brunswick: 1753. Not included in Readex Wheat/Brun: 161. Map of Maine; issued as a broadside by the Proprietors of Brunswick Township to show the "limits of the claim of the Plymouth Company" (Wheat/Brun).

1754

402 Johnston, Thomas (c. 1708-1767). *To His Excellency William Shirley, This Plan of Kennebeck and Sagadahock Rivers and Country Adjacent (Whereon Are Delineated the Boundaries of Several Ancient Grants), Being Taken from Actual Surveys Made by Joseph Heath Esq., Mr. Phineas Jones, John North Esqr., and Mr. Ephraim Jones is Inscribed.* Boston: Thomas Johnston, 1754. Not included in Readex Wheat/Brun: 163. Another map of Maine; according to Wheat and Brun, Johnston was reimbursed at least in part by the Plymouth Company.

403 Woodmason, Charles. *A Letter from a Gentleman of South Carolina on the Cultivation of Indigo.* [Charleston: 1754]. Bristol:1696 Not included in Readex. Apparently a discussion of raising indigo.

1755

404 *An Account of the Distances from the City of Philadelphia of All the Places of Note within the Improved Part of the Province of Pennsylvania.* Philadelphia: William Bradford, [1755]. Evans: 7345. Table of distances from the Philadelphia Court House to various destinations.

405 Evans, Lewis (1700-1756). *Geographical, Historical, Political, Philosophical and Mechanical Essays: The First Containing an Analysis of a General Map of the Middle British Colonies in America and of the Country of the Confederate Indians: A Description of the Face of the Country; the Boundaries of the Confederates; and the Maritime and Inland Navigations of the Several Rivers and Lakes Contained Therein.* Philadelphia: Franklin and Hall, 1755. Evans: 7411 Hazen: 3612 Meisel: 3: p. 346 Miller: 605. Collection of maps and geographical descriptions of the regions; it was written to encourage British settlement of the Ohio territory. Wheat and Brun describe the General Map as "one of the landmarks of American cartography" and state that the accuracy of the Ohio section is "remarkable for this period." The maps were republished and pirated many times in Europe, appearing as late as 1814 in a German edition.

406 Stevens, Thomas. *Method and Plain Process for Making Pot-Ash Equal If Not Superior to the Best Foreign Pot-Ash.* Boston: Edes and Gill, 1755. Evans: 7575 Not included in Readex. Stevens was English, but publicized his method in America. He succeeded in 1757 in convincing the Virginia legislature to build a potash furnace at Williamsburg.

1756

407 Clement, Timothy. *To His Excellency William Shirley Esqr., This Plan of Hudson's River from Albany to Fort Edward (and the Road from Thence to Lake George As Survey'd), Lake George, the Narrows, Crown Point, Part of Lake Champlain with Its South Bay, and Wood Creek, According to the Best Accounts from the French Genl's Plan and Other Observations (by Scale No. 1) and an Exact Plan of Fort Edward and William Henry (by Scale No. 2) and the West End of Lake George and of the Land Defended on the 8th of Sept. Last and of Our Army's Intrenchment Afterward (by Scale 3) and Sundry Particulars Respecting the Late Engagement with the Distance and Bearing of Crown Point and Wood Creek from No. 4, by Your Most Devoted Humble Servant Timo. Clement, Survr.* Boston: Thomas Johnston, 1756. Evans: 7390 Wheat/Brun: 322. Map of northern New York state.

408 Evans, Lewis (1700-1756). *Geographical, Historical, Political, Philosophical and Mechanical Essays, Number II.* Philadelphia: 1756. Evans: 7652. Evans' answer to those angered by Part 1 (see 405), chiefly the Shirley faction in New York. This map too was extensively pirated.

409 Fisher, Joshua (1707-1783). *Chart of Delaware Bay from the Sea Coast to Reedy Island.* [Philadelphia]: 1756. Evans: 7657 Wheat/Brun: 475. Navigation-style map, with soundings, shoals, and other features. The map was suppressed by Governor Morris because he feared it would aid the French; Fisher had the map re-engraved in a smaller size, covering the river around Philadelphia.

1757

410 Mascarene, John (1722-1779). *The Manufacture of Pot-Ash in the British North-American Plantations Recommended.* Boston: Z. Fowle, 1757. Evans: 7940 Hazen: 6813. Proposal to encourage potash manufacture as a way of increasing American exports to Britain. Mascarene makes suggestions for sites and pricing of potash, as well as projections as to the quantity to be produced.

1758

411 Gordon, John (b.1700). *John Gordon's Mathematical Traverse Table, &c.* Philadelphia: Dunlap; New York: Noel; Boston: Mecom, 1758. Evans: 8140 Karpinsky: p. 66. Trigonometry text intended for surveying and navigation; includes testimonials from Franklin and Aaron Burr, among others.

1759

412 Eliot, Jared (1685-1763). *The Sixth Essay on Field-Husbandry, As It Is, or May Be Ordered in New England.* New Haven: J. Parker, 1759. Evans: 8344. Devoted largely to the silk industry; it also includes arguments against slavery and the plantation system.

413 Scull, Nicholas (1787-1761). *To the Honorable Thomas Penn and Richard Penn, True and Absolute Proprietaries and Governours of the Province of Pennsylvania and Counties of New Castle, Kent, and Sussex on Delaware, This Map of the Improved Part of the Province of Pennsylvania Is Humbly Dedicated by Nicholas Scull.* Philadelphia: John Davis, 1759. Evans: 8489 Wheat/Brun: 422. Map of Pennsylvania as far north as the northeast branch of the Susquehanna and Upper Smithfield, as far south as Delaware; no details are shown on the New Jersey side of the Delaware River.

1760

414 Eliot, Jared (1685-1763). *Essays upon Field-Husbandry in New England As It Is or May Be Ordered.* Boston: Edes and Gill, 1760. Evans: 8590. Collection of all six of Eliot's essays on agriculture.

415 Gordon, John (b.1700). *A Supplement to the Mathematical Traverse Tables in Epitome.* Philadelphia: W. Dunlap, [1760]. mp: 41127 Bristol: 2133. Tables with brief explanations for their use (see 411).

1761

416 Milton, Abraham. *The Farmer's Companion, Directing How to Survey Land after a New and Particular Method.* Annapolis: 1761. mp: 41223 Bristol: 2231 Not included in Readex Karpinsky: p. 68. No copy is known.

1762

417 DeBruls, Michael. *Plan of Niagra [sic] with the Adjacent Country Surrendred [sic] to the English Army under the Command of Sr. Willm. Johnson, Bart. on the 25th of July, 1759.* [New York: 1762]. Not included in Readex Wheat/Brun: 324. Detailed map showing disposition of troops.

418 Eliot, Jared (1685-1763). *An Essay on the Invention or Art of Making Very Good, If Not the Best Iron from Black Sea Sand.* New York: John Holt, 1762. Evans: 9109 Hazen: 3475. Eliot's discovery of iron ore-bearing sea sand; he was awarded a gold medal by the London Society of the Arts for this work.

419 Scull, Nicholas (1687-1761). *To the Mayor, Recorder, Aldermen, Common Council and Freemen of Philadelphia, This Plan of the Improved Part of the City, Surveyed and Laid Down by the Late Nicholas Scull Esq., Surveyor General of the Province of Pennsylvania, Is Inscribed by the Editors.* Philadelphia: 1762. Evans: 9267 Wheat/Brun: 456. Wheat and Brun call this "the most important engraved plan of Philadelphia since the Thomas Holme plan of 1682."

1763

420 *A General Chart of All the Coast of the Province of Louisiana, the Bays of Ascension and Mobille [sic], Pensacola, and St. Joseph, with the Rivers Missisipi [sic] and Apalachy [sic], in the Gulph [sic] of Mexico.* Philadelphia: Matthew Clarkson, [c. 1763]. Not included in Readex Wheat/Brun: 630. According to Wheat and Brun, Clarkson advertised this map as "published from a draught taken by order of the French king."

1765

421 Quincy, Edmund (1703-1788). *A Treatise of Hemp Husbandry; Being a Collection of Approved Instructions As to the Choice and Preparation of the Soils Most Proper for the Growth of That Useful and Valuable Material, and Also As to the Subsequent Management Thereof, Agreeable to the Experience of Several Countries Wherein It Has Been Produced, Both in Europe and America.* Boston: Green and Russell, 1765. Evans: 10151. Argument for increased cultivation of hemp as cash crop; includes complete directions for cultivation and processing, as well as a diagram for a hemp mill. It was printed by the Massachusetts House of Representatives.

422 Wiley, John. *A Treatise on the Propagation of Sheep, the Manufacture of Wool, and the Cultivation and Manufacture of Flax, with Directions for Making Several Utensils for the Business.* Williamsburg: J. Royle, 1765. Evans: 10214 Guerra: a-376. Directions for care and breeding of sheep as well as processing of wool; includes a section on cultivating flax and processing linen. Wiley argues that wool and flax cultivation can help Virginia recover from a recession caused by falling tobacco prices.

1766

423 *Directions for Making Calcined or Pearl-Ashes, As Practised in Hungary, &c. with a Copper-Plate Drawing of a Calcined Furnace.* Boston: John Mein, 1766. Evans: 10285. Reprint of a letter to the Society for Promoting Arts and Sciences; urges production of "Pearl-Ashes" (a particularly pure form of potash) as an American industry. The pamphlet includes simple directions and a diagram.

424 Park, Moses. *To the Honourable the Earl of Shelburn [sic], His Majesty's Principal Secretary of State for the Southern Department, This Plan of the Colony of Connecticut in North-America Is Humbly Dedicated by His Lordship's Most Obedient Humble Servt. Moses Park, Novr. 24, 1766.* [New London?]: 1766. Not included in Readex Wheat/Brun: 257. Map prepared under the sponsorship of the Connecticut government; originally requested by the British in 1765 in an effort to improve American postal service.

1768

425 *Directions for Sailing in and out of Plymouth Harbour; Taken by Moses Bennet, William Rhodes, Thomas Allen, and Nathaniel Green, Being a Committee of the Marine Society of Boston Appointed for the Survey by the Desire of a Committee of the General Court of the Massachusetts Bay, Appointed to Build a Light-House on the Gurnet near Plymouth-Harbor in Said Province in July, 1768.* [Boston, 1768]. Evans: 10882. Navigation directions, with descriptions of landmarks, distances and soundings.

426 *A Plan of the Boundary Lines between the Province of Maryland and the Three Lower Counties of Delaware, with Part of the Parallel of Latitude Which Is the Boundary between the Provinces of Maryland and Pennsylvania.* [Philadelphia: Robert Kennedy, 1768]. Not included in Readex Wheat/Brun: 497. Part of the Mason/Dixon survey settling the boundary dispute between Pennsylvania and Maryland. Wheat and Brun call it "definitely one of the landmarks of American eighteenth century surveying."

427 *A Plan of the West Line or Parallel of Latitude, Which Is the Boundary between the Provinces of Maryland and Pansylvania [sic].* [Philadelphia: Robert Kennedy, 1768]. Not included in Readex Wheat/Brun: 498. The second part of the Mason/Dixon survey; originally published in three separate strips on one sheet.

428 Ratzer, Bernard. *To His Excellency Sir Henry Moore, Captain General and Governour in Chief in and over His Majesty's Province of New York and the Territories Depending Thereon in America, Chancellor and Vice Admiral of the Same, This Plan of the City of New York and Its Environs, Survey'd and Laid Down, Is Most Humbly Dedicated by His Excellency's Most Obedient and Humble Servant, B. Ratzer.* [New York, 1769]. Evans: 11434. Map of New York City.

1770
429 Scull, William (fl. 1765). *To the Honourable Thomas Penn and Richard Penn Esquires, True and Absolute Proprietaries and Governors of the Province of Pennsylvania and the Territories Thereunto Belonging, and to the Honourable John Penn Esquire, This Map of the Province of Pennsylvania Is Humbly Dedicated by Their Most Obedient Humble Servant William Scull.* Philadelphia: James Nevil, 1770. Evans: 11850 Wheat/ Brun: 425. Map of Pennsylvania.

1771
430 *Directions for the Breeding and Management of Silkworms, Extracted from the Treatises of the Abbe Boissier de Sauvages, and Pullein.* Philadelphia: Joseph Crukshank and Isaac Collins, 1771. Evans: 11574. Includes both directions from de Sauvages and Samuel Pullein and a review of petitions to the Pennsylvania legislature from the American Philosophical Society arguing for the establishment of a public "filature" with premiums for silk cocoons. The pamphlet also includes a letter from Franklin arguing for increased silk production. The appeal to the legislature was unsuccessful, but the filature was later built by private subscription .

431 Moody, Thomas (fl. 1771). *A Compendium of Surveying, or the Surveyor's Pocket Companion.* Burlington, NJ: Isaac Collins, 1771. Evans: 12129. Instructions in simple trigonometry supported by tables; also includes case studies, examples, and instructions in surveying techniques.

432 Ottolenghe, Joseph. *Directions for Breeding Silk-Worms, Extracted from a Letter of Joseph Ottolenghe, Esq., Late Superintendent of the Public Filature in Georgia.* Philadelphia: Joseph Crukshank, 1771. Evans: 12172. Instructions for breeding and growing silk worms, as well as processing the silk.

1774

433 Carter, John. *The Young Surveyor's Instructor.* Philadelphia: W. and T. Bradford, 1774. mp: 42570 Bristol: 3709 Karpinsky: p. 79. Textbook of surveying, with tables.

434 *Directions for the Gulph [sic] and River of St. Lawrence with Some Occasional Remarks.* Philadelphia: William and Thomas Bradford, 1774. Evans: 13250. Directions for navigating the St. Lawrence with distances, soundings, and descriptions of landmarks.

435 Reed, John. *To the Honourable House of Representatives of the Freemen of Pennsylvania, This Map of the City and Liberties of Philadelphia with the Catalogue of Purchasers Is Humbly Dedicated by Their Most Obedient Humble Servant John Reed.* Philadelphia: Thomas Man, [1774]. Evans: 13564 Wheat/Brun: 457. Shows the streets and squares as originally laid out by Holme and advertised by Penn; includes a list of the original purchasers with their lots. There is an inset map showing the newer street layout.

1775

436 *An Essay on the Culture and Management of Hemp, More Particularly for the Purpose of Making Coarse Linens.* Annapolis: Anne Catherine Green, 1775. Evans: 14022. Description of cultivation, harvesting and processing of hemp; includes arguments for hemp cultivation rather than flax. The pamphlet was produced for the purpose of stimulating American textile production to make up for blocked European imports.

437 Loocock, Aaron. *Some Observations and Directions for the Culture of Madder.* [Charles-Town: Peter Timothy, 1775]. Evans: 14166. Description of cultivation and processing of madder, a plant used in red dye.

438 Romans, Bernard (c.1720-c.1784). *Chart Containing Part of East Florida, the Whole Coast of West Florida with All the Soundings &c., All the Mouths of the Mississippi.* [New York, 1775]. Evans: 14441 Wheat/Brun: 621. Wheat and Brun call it Romans' "famous chart of East and West Florida"; based on Romans' surveying work in Florida. The actual engraving of the map was done by Paul Revere and Abel Buell.

439 Romans, Bernard (c.1720-c.1784). *Chart Containing the Peninsula of Florida, the Bahama Islands, the North Side of the Island of Cuba, the Old Streight [sic] of Bahama, and All the Islands, Keys, Rocks, &c. in These Seas.* [New York, 1775]. Evans: 14442. Not included in Readex. Map of Florida and Caribbean.

440 Romans, Bernard (c.1720-c.1784). *A Concise Natural History of East and West Florida; Containing an Account of the Natural Produce of All the Southern Part of British America in the Three*

Kingdoms of Nature, Particularly the Animal and Vegetable. New York: 1775. Evans: 14440 Guerra: a-578 Hazen: 9012 Meisel: 3: p. 350. Geography of Florida; Romans' most famous book. It was published after Romans had been surveyor of the Southern District, exploring Florida intermittently from 1766-1771. A second volume was proposed, but not published.

441   Romans, Bernard (c.1720-c.1784). *To the Hon'l Jno. Hancock Esqre., President of the Continental Congress, This Map of the Seat of Civil War in America Is Respectfully Inscribed by His Most Obedient Servant, B. Romans.* [Philadelphia, 1775]. Evans: 14444 Wheat/Brun: 203. Maps of Massachusetts and Rhode Island.

442   United States Continental Congress. *Several Methods of Making Salt-Petre; Recommended to the Inhabitants of the United Colonies by their Representatives in Congress.* Philadelphia: W. and T. Bradford, 1775. Evans: 14584. Contains both recommendations for the manufacture of saltpeter (necessary for gun powder) and explanations of the process. The pamphlet is taken almost entirely from articles by Rush and Franklin.

1776

443 *A Map of Forty Miles North, Thirty Miles West, and Twenty-Five Miles South of Boston, Including an Accurate Draft of the Harbor and Town.* n.p.: [c.1776]. Wheat/Brun: 205 Not included in Readex. Map apparently dating from immediately after the Bunker Hill, Concord, and Lexington battles; all three are shown on the map.

444   New York (State) Committee of Safety. *Essays upon the Making of Salt-Petre and Gun-Powder.* New York: Samuel Loudon, 1776. Evans: 14930. Includes material included in 442 and 445, as well as new material and the Salt Petre Act passed by the Continental Congress.

445   [Paine, Robert Treat (1731-1814)]. *The Art of Making Common Salt, Particularly Adapted to the Use of the American Colonies, with an Extract from Dr. Brownrigg's Treatise on the Art of Making Bay-Salt.* Philadelphia: R. Aitken, 1776. Evans: 14651 Hazen: 1319. Step-by-step process description with diagrams. Paine was a member of the Continental Congress and a signer of the Declaration of Independence; he was a co-founder of the American Academy of Arts and Sciences in 1780.

446   Philadelphia Committee of Safety. *The Process for Extracting and Refining Salt-Petre, According to the Method Practised at the Provincial Works in Philadelphia.* Philadelphia: William and Thomas Bradford, 1776. Evans: 14827. Another description of the salt peter manufacturing process.

447 Turner, James (d. 1759). *Map of the Province of Nova Scotia and Parts Adjacent.* Philadelphia: R. Aitken, 1776. 4th edition. Evans: 15121 Wheat/Brun: 92. First published by Turner in Boston, c. 1750.

1777

448 [Romans, Bernard (c.1720-c.1784)]. *Connecticut and Parts Adjacent.* New Haven: [1777]. Evans: 15585 Not included in Readex Wheat/Brun: 261. Wheat and Brun term it a "political and topographical map"; it is attributed to Romans on the basis of newspaper advertisements.

449 *Select Essays on Raising and Dressing Flax and Hemp and on Bleaching Linen-Cloth, with Valuable Dissertations on Other Useful Subjects.* Philadelphia: Robert Bell, 1777. Evans: 15597. Collection of agricultural advice from various sources (chiefly European); besides flax and hemp, it covers paper making, cattle and sheep raising, and vegetable cultivation. It includes diagrams of an apparatus for preparing flax.

1778

450 Churchman, John. *To the American Philosophical Society, This Map of the Peninsula between Delaware and Chesopeak [sic] Bays with the Said Bays and Shores Adjacent, Drawn from the Most Accurate Surveys, Is Inscribed by John Churchman.* [Baltimore: 1778?]. mp: 44871 Bristol: 6264 Wheat/Brun: 477. Map showing roads, streams and rivers in the Chesapeake region; includes proposed routes for five canals.

451 *Directions to Sail into and up Delaware Bay.* [Philadelphia, 1778]. Evans: 15784. Navigation directions including soundings and description of natural landmarks.

452 Romans, Bernard (c.1720-c.1784). *A Chorographical Map of the Country round Philadelphia.* n.p.: [1778]. Not included in Readex Wheat/Brun: 304. Includes Valley Forge and the "tracks" of the armies of Washington and Howe; drawn and engraved by Romans.

453 [Romans, Bernard (c.1720-c.1784)]. *The Townships or Grants East of Lake Champlain. . . .* [New Haven: 1778]. mp: 43551 Bristol: 4791. Map of northern New York.

1779

454 *A Chorographical Map of the Northern Department of North America, Drawn from the Latest and Most Accurate Observations.* [New Haven: Amos Doolittle,1779]. mp: 43609 Bristol: 4851 Wheat/Brun: 147. Similar to map published by Samuel Holland in London, 1775 covering New York, New Jersey, and Pennsylvania.

1780

455 *A Map of Tottin [sic] and Crosfields [sic] Purchase and the Waters Adjacent in the State of New York.* n.p.: [1780?]. Not included in Readex Wheat/Brun: 329. Map of land purchased by Joseph Totten and Stephen Crossfield in 1771.

456 Matlack, Timothy (d. 1829). *An Oration Delivered March 16, 1780, before the Patron, Vice Presidents, and Members of the American Philosophical Society Held at Philadelphia for Promoting Useful Knowledge.* Philadelphia: Styner and Cist, 1780. Evans: 16867. Speech on the rise, progress and early history of agriculture, with some consideration of its present state. This was the first of the annual American Philosophical Society orations delivered after a five year hiatus occasioned by the Revolution. Matlack claimed, with what Hindle calls "patriotic zeal," that the French were responsible for the agricultural revolution.

457 *Plan of Part of the District of Main.* n.p.: [c.1780]. Not included in Readex Wheat/Brun: 164. Map of various surveyed ranges and townships in Maine.

1782

458 Bauman, Sebastian. *To His Excellency General Washington, Commander in Chief of the Armies of the United States of America, This Plan of the Investment of York and Gloucester Has Been Surveyed and Laid Down, and is Most Humbly Dedicated by His Excellency's Obedient and Very Humble Servant, Sebastn. Bauman, Major of the New York or 2nd Regt. of Artillery.* Philadelphia: 1782. mp: 44173 Bristol: 5490 Wheat/Brun: 541. Map of the battlefield, prepared in the week following Cornwallis' surrender. Many reduced copies were later published.

459 Dearborn, Benjamin (1755-1838). *The Pupil's Guide.* Portsmouth, NH: Daniel Fowle, 1782. Evans: 17510 Karpinsky: p. 82. Arithmetic text; chiefly a list of rules and maxims.

1783

460 Buell, Able (1750-1825). *A New and Correct Map of the United States of North America.* New Haven: [1783]. Evans: 19533 Wheat/Brun: 109-10. Map taken largely from John Mitchell and Lewis Evans. Wheat and Brun call it "an important landmark in the history of American engraving." It was the first map of the United States according to the Peace of 1783 drawn by an American; it also has the first printed version of the American flag drawn in the cartouche.

1784

461 [Bordley, John Beale (1727-1804)]. *A Summary View of the Courses of Crops in the Husbandry of England and Maryland; with a Comparison of Their Products and a System of Improved Courses*

Proposed for Farms in America. Philadelphia: Charles Cist, 1784.
Evans: 18373. Discussion of crop rotation, including diagrams of
planting and rotation over a field during four years. Based on English
models.

462 **[Davidson, Robert (1750-1812)].** *Geography Epitomized, or a
Tour round the World; Being a Short but Comprehensive Description of
the Terraqueous Globe, Attempted in Verse (for the Sake of the
Memory) and Principally Designed for the Use of Schools.*
Philadelphia: Joseph Crukshank, 1784. Evans: 18435. Popular for its
"ingenious rhymes"; reprinted in London and Burlington, New Jersey.
Davidson was professor of geography, history, chronology, rhetoric
and belles-lettres at Dickinson College in 1784.

463 **Filson, John (c.1747-1788).** *The Discovery, Settlement, and
Present State of Kentucke [sic]: an Essay towards the Topography, and
Natural History of That Important Country.* Wilmington: James
Adams, 1784. Evans: 18467 Hazen: 3822 Meisel: 3: p.353. First
history of Kentucky, including the first map and the first account of
Daniel Boone; intended to encourage settlement. According to Wheat
and Brun the map is "one of the best known of early American printed
maps" and is considered relatively accurate.

464 *Geographical Gazetteer of the Towns in the Commonwealth of
Massachusetts.* [Boston:1784]. mp: 44535 Bristol: 5899. Description
of towns of Massachusetts.

465 **Hutchins, Thomas (1730-1789).** *An Historical Narrative and
Topographical Description of Louisiana and West Florida.* Philadelphia:
Robert Aitken, 1784. Evans: 18532 Hazen: 5493. Based on reports
and descriptions compiled by Hutchins when he was with the British
army in West Florida (1772-76); Aitken was unable to raise the money
to print the map originally intended to accompany the book.

466 **McMurray, William.** *The United States According to the
Definitive Treaty of Peace Signed at Paris, Septr. 3d 1783.*
[Philadelphia: c.1784]. Evans: 20476 Wheat/Brun: 111. Map of the
country from the Mississippi to the Atlantic, north to the Canadian
border and south to South Carolina.

467 **Mendenhall, Thomas (1759-1843).** *Traverse Tables or Tables
of Difference of Latitude and Departure, Constructed to Every Quarter
of a Degree of the Quadrant, and to Answer for All Distances from One
Tenth of a Perch or Mile to Twelve Hundred.* Philadelphia: Joseph
Crukshank, 1784. Evans: 18609. Tables for surveyors with
instructions for their use; includes endorsements by David Rittenhouse,
John Lukens, and Thomas Hutchins.

468 Morse, Jedidiah (1761-1826). *Geography Made Easy.* New Haven: Meigs, Bowen and Dana, [1784]. Evans: 18615. Geography textbook consisting of a series of lectures Morse delivered at a New Haven girls' school; the "first geography to be published in the United States" (DAB). It was extremely popular, going through 25 editions; its success convinced Morse to become a geographer.

469 Parker, Abner. *Captain Parker's Chart of Saybrook Bar.* n.p.: [Abel Buel], [c. 1784]. Not included in Readex Wheat/Brun: 264. Chart of the mouth of the Connecticut River; according to Wheat and Brun it was part of an effort by the river communities to make the river more accessible from Long Island Sound.

1785
470 [Colles, Christopher (1738-1816)]. *Proposals for the Speedy Settlement of the Waste and Unappropriated Lands on the Western Frontiers of the State of New York and for the Improvement of the Inland Navigation between Albany and Oswego.* New York: Samuel Loudon, 1785. Evans: 18960. Proposals for development of 250,000 acres of land in western New York and for building a canal from Albany to Lake Ontario with Colles as engineer. Hindle calls it a "magnificent proposal."

471 Gibson, Robert. *A Treatise on Practical Surveying, Which Is Demonstrated from Its First Principles.* Philadelphia: Joseph Crukshank, 1785. Evans: 19026 Karpinsky: p. 83. Surveying textbook; covers basic geometry, trigonometry, and the use of surveying instruments and techniques.

472 McDonald, Alexander (c.1752-1792). *The Youth's Assistant; Being a Plain, Easy, and Comprehensive Guide to Practical Arithmetic: Containing All the Rules and Examples Necessary for Such a Work.* Norwich: John Trumbull, 1785. Evans: 19066 Karpinsky: p. 85. Arithmetic textbook.

473 Norman, John (1748-1817). *A New Map of the New England States.* [Boston: Norman and Coles, 1785]. Evans: 19144 Not included in Readex Wheat/Brun: 148. According to Wheat and Brun, the "only probable copy" of this map has a section missing from the lower right corner that probably contained the title imprint.

474 Philadelphia Society for Promoting Agriculture. *An Address from the Philadelphia Society for Promoting Agriculture; with a Summary of Its Laws and Premiums Offered.* [Philadelphia]: 1785. Evans: 19193. Description of the Society's purpose, with a comparison of American and European agriculture. The "premiums" (cash awards) were offered by the Society to farmers undertaking trials of various European agricultural methods.

475   South Carolina Society for Promoting and Improving Agriculture. *Address and Rules of the South Carolina Society for Promoting and Improving Agriculture and Other Rural Concerns.* Charleston: A. Timothy, 1785. Evans: 19254 Hazen: 9931. Proposal that South Carolina planters experiment in agricultural methods; includes rules of the Society.

476   *A Synopsis of Geography, with the Use of the Terrestrial Globe.* Wilmington: James Adams, 1785. mp: 44796 Bristol: 6176. Grammar school geography textbook.

477   Varlo, Charles (1725-1795). *A New System of Husbandry.* Philadelphia: 1785. Evans: 19338. An agricultural textbook with advice on crop cultivation, raising livestock, and processing madder, hemp, and flax. It closes with a monthly calendar of farming chores and includes a review of Newtonian theories of light and plant anatomy. Varlo was an English author, but he includes an introduction in which he claims American experience.

1786

478   Bowler, Metcalf (1726-1789). *A Treatise on Agriculture and Practical Husbandry.* Providence: Bennett Wheeler, 1786. Evans: 19522. Scientific treatment of agriculture, asserting that agriculture "is a material branch of experimental philosophy." According to Hindle, the work "added little that was novel but . . . did prove a popular presentation of the new agriculture."

479   Fitch, John (1743-1798). *A Map of the Northwest Parts of the United States of America.* Philadelphia: Dobson, 1786. Evans: 19648 Wheat/Brun: 660. Map of the region from the Great Lakes south to North Carolina; based on maps by McMurray and Hutchins. According to Wheat and Brun it is the only American map of the period to be made, engraved, and printed by the same person.

480   Franklin, Benjamin (1706-1790). *Maritime Observations in a Letter from Doctor Franklin to Mr. Alphonsus Le Roy, Member of Several Academies at Paris.* Philadelphia: Robert Aitken, 1786. mp: 44888 Bristol: 6285. Suggestions for improving ship design and navigation techniques; it includes Franklin's observations on water temperature as well as drawings of his design suggestions.

481   [Norman, John (c.1748-1817)]. *The Town and Country Builder's Assistant; Absolutely Necessary to Be Understood by Builders and Workmen in General.* Boston: John Norman, [1786]. Evans: 20027. Simplified architecture text; describes various building elements. Most of the book is given over to diagrams and plans. Norman described himself as an "architect-engraver" in his business advertisements.

1787

482   Coxe, Tench (1755-1824). *An Address to an Assembly of the Friends of American Manufactures.* Philadelphia: R. Aitken and Son, 1787. Evans: 20305. Proposal for establishing a society for the encouragement of American manufacturing; Coxe was a fervent supporter of American industry and a president of the Pennsylvania Society for the Encouragement of Manufactures and the Useful Arts.

483   [Cutler, Manasseh (1742-1823)]. *An Explanation of the Map Which Delineates That Part of the Federal Lands Comprehended between Pennsylvania West Line, the Rivers Ohio and Sioto, and Lake Erie; Confirmed to the United States by Sundry Tribes of Indians, in the Treaties of 1784 and 1785, and Now Ready for Settlement.* Salem: Dabney and Cushing, 1787. Evans: 20312 Wheat/Brun: 662. First map showing the "Seven Ranges of Townships" in the Ohio Valley. Cutler was a member of the Ohio Company whose aim was the colonization of the valley; in 1787 he became the Company's agent to Congress. The map and explanation were sold separately.

484   Hutchins, Thomas (1730-1789). *A Topographical Description of Virginia, Pennsylvania, Maryland, and North Carolina, Comprehending the Rivers Ohio, Kenhawa, Sioto, Cherokee, Wabash, Illinois, Mississippi, &c.* Boston: John Norman, 1787. Evans: 20424 Hazen: 5495 Meisel: 3: p. 352. Geography of the region; includes appendix with Patrick Kennedy's journal of travels in Illinois. Hutchins was "Geographer to the United States," charged with the surveying of the Ohio Valley under the Land Ordinance of 1785.

485 *Notes on Farming.* New York: 1787. Evans: 20599. Miscellaneous farming advice including crop cultivation, raising livestock, and managing an orchard.

486   Pennsylvania Society for Encouragement of Manufactures. *The Plan of the Pennsylvania Society for Encouragement of Manufactures and the Useful Arts.* Philadelphia: R. Aitken, 1787. Evans: 20637. Charter for the society using some of Coxe's suggestions (see 482); it establishes a "manufacturing fund" to back factories in suitable locations.

487 *Simple Division.* Boston: James White, [1787]. mp: 45164 Bristol: 6590. Broadside explaining rules of division.

488   Squibb, Robert. *The Gardener's Calendar for South Carolina, Georgia, and North Carolina.* Charleston: Samuel Wright, 1787. Evans: 20722. Description of work necessary in "kitchen and fruit gardens" in monthly form. It includes directions for growing various plants and trees native to the area; Squibb was supported by the South Carolina Society for Promoting and Improving Agriculture.

1788

489 American Philosophical Society. *At a Meeting of the American Philosophical Society.* Philadelphia: 1788. mp: 45218 Bristol: 6660. Brief acknowledgement of James Rumsey's plans for a steam engine and improvements of saw mill and grist mill design (see 499); includes a cautious approval of Rumsey's theories.

490 Barnes, Joseph. *Remarks on Mr. John Fitch's Reply to Mr. James Rumsey's Pamphlet.* Philadelphia: Joseph James, 1788. Evans: 20954. Defense of Rumsey as originator of steam navigation (Barnes was Rumsey's assistant); includes documents supporting Rumsey's claims in an appendix (see 491 and 500).

491 [Fitch, John (1743-1798)]. *The Original Steam-Boat Supported, or a Reply to Mr. James Rumsey's Pamphlet.* Philadelphia: Zachariah Poulson, 1788. Evans: 21092. Fitch's claim to have invented a steamboat prior to Rumsey (see 500); includes supporting documents and Rumsey's original pamphlet. Fitch maintained the exclusive right to sail steamboats on all waters of Pennsylvania, New York, Delaware, and Virginia.

492 Gough, John (1721-1791) and Benjamin Workman. *A Treatise of Arithmetic in Theory and Practice, to Which Are Added Many Valuable Additions and Amendments, More Particularly Fitting the Work for the Improvement of the American Youth by Benjamin Workman.* Philadelphia: J. McCulloch, 1788. Evans: 21116 Karpinsky: p. 89. Originally a British text, with American revisions by Workman including the use of American currency in examples. The book had two more editions and a "continuation" in 1798 with the addition of a short section on algebra.

493 Jefferson, Thomas (1743-1826). *Notes on the State of Virginia.* Philadelphia: Prichard and Hall, 1788. Evans: 21176 Meisel: 3: p. 352. Jefferson's influential description, including geography and natural history. It was originally published in France based on material assembled by Jefferson during and after his term as governor of Virginia. Jefferson submitted the original manuscript for criticism and revision to Thomas Walker, George Rogers Clark, and Thomas Hutchins among others. Many editions include a map drawn by Jefferson based on one drawn by his father and Joshua Fry in 1741.

494 Minto, Walter (1753-1796). *An Inaugural Oration on the Progress and Importance of the Mathematical Sciences.* Trenton: Isaac Collins, 1788. Evans: 21260 Karpinsky: p. 89. Brief summary of the history of mathematics; according to Karpinsky, Minto shows "remarkable familiarity with the most important names." Minto was professor of mathematics at the College of New Jersey (Princeton); the oration was delivered on the evening before the annual commencement.

495   Pike, Nicholas (1743-1819). *A New and Complete System of Arithmetic, Composed for the Use of the Citizens of the United States.* Newburyport: John Mycall, 1788. Evans: 21394 Karpinsky: p. 90. Very popular arithmetic text endorsed by mathematician Benjamin West and the presidents of Yale, Harvard and Dartmouth. The DAB calls it "an admirable effort, furnishing excellent material in geometry and trigonometry." There were eight more editions.

496   Pinkham, Paul and Alex Coffin. *Directions to and from the Light-House on the Northeast Point of Nantucket.* [Boston: 1788]. Evans: 21058. Navigation directions with distances from landmarks.

497   Purcell, Joseph. *A Map of the States of Virginia, North Carolina, South Carolina, and Georgia, Comprehending the Spanish Provinces of East and West Florida, Exhibiting the Boundaries As Fixed by the Late Treaty of Peace between the United States and the Spanish Dominions.* New Haven: 1788. Evans: 21412 Wheat/Brun: 491. Detailed map showing the boundaries claimed by the Indian Nations and the trading paths.

498   Ruggles, Edward Jr. (fl. 1789-1800). *A New Map of the World.* Pomfret, CT: 1788. Evans: 21213 Not included in Readex Wheat/Brun: 5. Includes late discoveries, particularly the track of Cook's last voyage.

499   Rumsey, James (1743-1792). *The Explanations and Annexed Plates of the Following Improvements in Mechanics, Viz 1st, a New Constructed Boiler, 2nd, a Machine for Raising Water, 3d, a Grist-Mill, 4th, a Saw-Mill, Are Respectfully Dedicated to the Patrons of the Plans by the Inventor. Explanation of a Steam Engine, and the Method of Applying It to Propel a Boat.* Philadelphia: Joseph James, 1788. Evans: 21439. Two separate pamphlets bound together; they feature a series of engraved diagrams with short explanations of each design.

500   Rumsey, James (1743-1792). *A Short Treatise on the Application of Steam, Whereby It Is Clearly Shewn, from Actual Experiments, That Steam May Be Applied to Propel Boats or Vessels of Any Burthen against Rapid Currents with Great Velocity.* Philadelphia: Joseph James, 1788. Evans: 21441. Rumsey's descriptions of his steamboat design and his account of his dispute with John Fitch (see 491) in which he unsuccessfully challenged Fitch's monopoly on steam navigation.

501   Wall, George, Jr. (fl. 1788). *A Description with Instructions for the Use of a Newly Invented Surveying Instrument, Called the Trigonometer.* Philadelphia: Zachariah Poulson, 1788. Evans: 21568 Karpinsky: p. 92. Description of surveying instrument with instructions for use, diagrams, and patent information.

502 Workman, Benjamin. *Gauging Epitomized.* Philadelphia: W. Young, 1788. Evans: 21618. Method for estimating the contents of objects such as ships and casks; includes tables, examples, and calculations.

1789

503 Blodget, William. *A Topographical Map of Vermont, from Actual Survey.* New Haven: Amos Doolittle, 1789. Evans: 21696 Wheat/Brun: 192. Shows county and township boundaries as well as many place names.

504 [Bordley, John Beale (1727-1804)]. *On Monies, Coins, Weights, and Measures Proposed for the United States of America.* Philadelphia: Daniel Humphreys, 1789. Evans: 21698. Explanation and defense of decimal currency with gold and silver equivalents; includes an argument for decimal weights.

505 [Bordley, John Beale (1727-1804)]. *Purport of a Letter on Sheep.* [Philadelphia: 1789]. Evans: 21699. Description of a method of "sheep husbandry."

506 Clark, Matthew. *[A Complete Chart of the Coast of America].* Boston: Clark and Carleton, [1789]. Evans: 21738 Wheat/Brun: 93-96. Set of charts without title (title supplied by Evans); includes navigation directions and maps of the coastline from Florida northward.

507 Colles Christopher (1738-1816). *A Survey of the Roads of the United States of America.* [New York: 1789]. Evans: 21741. Series of maps of roads, chiefly in Pennsylvania and New York. The maps included accurate distances between cities measured by Colles' "mileometer," which he attached to a carriage wheel.

508 [Coxe, Tench (1755-1824)]. *Observations on the Agriculture, Manufactures and Commerce of the United States.* New York: Francis Childs and John Swaint, 1789. Evans: 21774 Hazen: 2678. Discussion of state of American agriculture and manufacturing directed to Congress. Although Coxe was a zealous promoter of industry, he admitted that American agriculture needed greater attention.

509 Morse, Jedidiah (1761-1826). *The American Geography, or a View of the Present Situation of the United States of America.* Elizabethtown: Shepard Kollock, 1789. Evans: 21978 Hazen: 7447. Geography text inspired by the success of *Geography Made Easy* (see 468); it was extremely popular, going through seven American editions and several European ones. Morse received advice from Jeremy Belknap, Samuel L. Mitchill, Thomas Hutchins, Noah Webster, and Ebenezer Hazard, among other; his maps were drawn by Joseph Purcell (southern states) and Amos Doolittle (northern states). The geography is largely derived from Thomas Hutchins and earlier works, such as those by Lewis Evans and Thomas Pownall.

510 Providence Association of Mechanics and Manufacturers. *The Charter, Constitution and Bye-Laws [sic] of the Providence Association of Mechanics and Manufacturers.* Providence: Bennett Wheeler, [1789]. Evans: 22085. Society charter and rules.

511 Ruggles, Edward, Jr. (fl. 1789-1800). *Plan of the City Marietta at the Confluence of the Rivers Ohio and Muskingum.* [New London:1789]. Evans: 22121 Wheat/Brun: 670. Map of Marietta with a brief description of lot size and the location of the city.

512 Sarjeant, Thomas. *An Introduction to the Counting House, or a Short Specimen of Mercantile Precedents, Adapted to the Present Situation of the Trade and Commerce of the United States of America.* Philadelphia: Dobson and Lang, 1789. Evans: 22127 Karpinsky: p. 92. Bookkeeping textbook.

513 Workman, Benjamin. *The American Accountant or School Master's New Assistant.* Philadelphia: John McCulloch, 1789. Evans: 22290 Karpinsky: p. 94. According to Karpinsky, this text is based on Gough's *Treatise of Arithmetic in Theory and Practice* which Workman had revised (see 492). There were two more editions of this work with revisions by Robert Patterson.

1790
514 Churchman, John (1753-1805). *An Explanation of the Magnetic Atlas or Variation Chart.* Philadelphia: James and Johnson, 1790. Evans: 22406 Hazen: 2370. Description of an improved system of finding longitude through tables of magnetic variations based on Churchman's theory of satellites revolving the earth at the poles. According to Hindle, the American Philosophical Society dismissed Churchman's theories as "groundless . . . whimsical . . . and impossible in the nature of things." Churchman subsequently submitted his theories to several other critics and attempted to raise funds from Congress for an expedition to Baffin Bay to prove his hypotheses.

515 Deane, Samuel (1733-1814). *The New England Farmer or Georgical Dictionary: Containing a Compendious Account of the Ways and Methods in Which the Most Important Art of Husbandry, in All Its Various Branches, Is, or May Be, Practised to the Greatest Advantage in This Country.* Worcester, MA: Isaiah Thomas, 1790. Evans: 22450 Meisel: 3: p. 356. Agricultural encyclopedia sponsored by the American Academy of Arts and Sciences. It contains both Deane's own work and material on other authors' observations and experiments gathered from Deane's reading.

516 *An Essay on the Culture of Silk and Raising White Mulberry Trees, the Leaves of Which Are the Only Proper Food of the Silk-Worm.* Philadelphia: Joseph Crukshank, 1790. Evans: 22491. Proposal for establishing a silk industry in the Middle Atlantic states; includes some description of mulberry cultivation.

517 **[Jefferson, Thomas (1743-1826)].** *Report of the Secretary of State, on the Subject of Establishing a Uniformity in the Weights, Measures and Coins of the United States.* New York: Francis Childs and John Swaine, 1790. Evans: 22994. Jefferson's response as Secretary of State to a request from the House of Representatives for a plan for establishing uniformity in weights and measures. Jefferson proposed standardizing the existing system or using a decimal system similar to that used for coinage. No action was taken on the report. A postscript to another edition in 1791 (Evans: 23910) explains how weights and measures were to be derived.

518 **Poellnitz, F.C.H.B.** *Essay on Agriculture.* New York: Francis Childs and John Swaine, 1790. Evans: 22805. Description of four agricultural implements (sowing plough, threshing mill, rippling cart, horse hoe), with diagrams and directions for use.

519 *Remarks on the Manufacturing of Maple Sugar with Directions for Its Further Improvement.* Philadelphia: James and Johnson, 1790. Evans: 22832. Description of process of obtaining maple sugar along with recommendations for increasing the maple sugar industry and extending it into the Middle Atlantic states.

520 **[Robertson, John (1712-1796)].** *Tables of Difference of Latitude and Departure, Constructed to Every Quarter of a Degree of the Quadrant, and Continued from One to the Distance of One Hundred Miles or Chains.* Philadelphia: Joseph Crukshank, 1790. Evans: 22856 Karpinsky: p. 126. Tables without explanation; usually bound with logarithms on the last 60 pages.

521 **Sterry, Consider (1761-1817), and John (1766-1823).** *The American Youth, Being a New and Complete Course of Introductory Mathematics Designed for the Use of Private Students.* Providence: Bennett Wheeler, 1790. Evans: 22910 Karpinsky: p. 95. Arithmetic text covering basic arithmetic and algebra; particular emphasis on commercial arithmetic.

522 *To the Patrons of the Columbian Magazine, This Map of Pennsylvania Is Dedicated by Their Obliged and Obed't Servants, the Proprietors.* [Philadelphia: William Spotswood, 1790]. Evans: 22804 Wheat/Brun: 431. Map of Pennsylvania.

1791

523 *An Address to the Manufacturers of Pot and Pearl Ash, with an Explanation of Samuel Hopkins's Patent Method of Making the Same.* New York: Childs and Swaine, 1791. mp: 46109 Bristol: 7629. Description of the manufacturing process with diagrams of a furnace.

524 Adlum, John (1759-1836) and John Wallis. *A Map Exhibiting a General View of the Roads and Inland Navigation of Pennsylvania and Part of the Adjacent States.* Philadelphia: 1791. Evans: 23104 Wheat/Brun: 432. Includes proposed roads and canals as well as existing roads, trails, and canals; also includes a description of the action of the canal locks.

525 Croswell, William (1760-1834). *Tables for Readily Computing the Longitude by the Lunar Observations.* Boston: I. Thomas and E. T. Andrews, 1791. Evans: 23300. Tables along with an explanation of their use.

526 Fraser, Donald (fl. 1797). *The Young Gentleman and Lady's Assistant, Partly Original, but Chiefly Compiled from the Works of the Most Celebrated Modern Authors; Calculated to Instruct Youth in the Principles of Useful Knowledge.* New York: Thomas Greenleaf, 1791. Evans: 23387 Karpinsky: p. 95. Hazen: 4099. Omnibus textbook including geography, "natural history" (meteorology, chemistry, geology, zoology), elocution, poetry, bible study, and "A Concise System of Practical Arithmetic." There were two more editions in 1794 and 1796.

527 Howell, Reading. *A Map of the State of Pennsylvania.* Philadelphia: 1791. Evans: 23454 Wheat/Brun: 433. Wheat and Brun term this "the best map of Pennsylvania to appear in the eighteenth century, and the first detailed map of the state to show its exact boundaries." The map was published from four plates, and several different versions of each plate were issued in succeeding years.

528 New York (City) Tammany Society. *American Museum under the Patronage of the Tammany Society or Columbian Order.* New York: Thomas and James Swords, [1791]. Evans: 23619. Announcement of the intention to found the American Museum to preserve objects related to American history and "all American curiosities of nature or art"; includes organization by-laws.

529 Norman, John (1748-1817). *The American Pilot, Containing the Navigation of the Sea Coasts of North America, from the Streights of Belle-Isle to Cayenne, Including the Island and Banks of Newfoundland, the West-India Islands, and All the Islands on the Coast.* Boston: John Norman, [1791]. Evans: 23637. Collection of navigation instructions and charts.

530   Norman, John (1748-1817). *The United States of America, Laid Down from the Best Authorities Agreeable to the Peace of 1783.* Boston: J. Norman, 1791. Evans: 23250 Not included in Readex Wheat/Brun: 119. Includes many soundings, surveyed townships and the seven ranges of the Ohio Valley; another edition was published in 1797 (Wheat/Brun:138) with new names added to the western portion..

531 Pennsylvania. *Copy of a Report from Reading Howell, Frederick Antes, and William Dean, Esquires, Commissioners Appointed to Explore the Head-Waters of the Rivers Delaware, Lehigh, and Schuylkill, and the Northeast Branch of Susquehanna.* Philadelphia: Francis Bailey, 1791. Evans: 23678. Description of roads and terrain in the region with suggestions for improvements in waterways; includes a budget for opening new waterways inland. The report was compiled by Timothy Matlack, Samuel Maclay, and John Adlum.

532 Pennsylvania. *Reports of Sundry Commissioners Appointed to View and Explore the Rivers Susquehanna and Juniata, the River Delaware, and the River Schuylkill, &c.* Philadelphia: Francis Bailey, 1791. Evans: 23679. Description of terrain, roads and waterways; includes suggestions for locks and canals. The authors included Reading Howell, Benjamin Rittenhouse, John Adlum, Timothy Matlack, and William Dean.

533 Tatham, William (1752-1819). *A Topographical Analysis of the Counties of the Commonwealth of Virginia, Compiled for the Years 1790-1.* Richmond: Thomas Nicolson, [1791]. Evans: 23820. Includes "extent and relative situation" of counties, origin of their names, bodies of water, distances, county officials, size of militia, court information, and population. Tatham organized the geographical department for Virginia.

1792
534 Belknap, Jeremy (1744-1798). *The History of New Hampshire, Vol. III.* Boston: Belknap and Young, 1792. Evans: 24088 Hazen: 1617 Meisel: 3: p. 353. "Geographical description" of New Hampshire, including physical characteristics, production, natural history, climate, commerce, and culture. Belknap interviewed hunters, surveyors, and scouts and circulated a letter of inquiry to the New Hampshire clergy; according to Greene he also acknowledged the contributions of Manasseh Cutler and William D. Peck in identifying local plants and animals.

535 Blodget, William. *A New and Correct Map of Connecticut, One of the United States of North America, from Actual Survey.* Middletown: 1792. Evans: 24124 Wheat/Brun: 273. Shows many buildings such as churches, academies, distilleries, etc.; also includes several new streams. However, Blodget omits some towns included by Romans (see 448).

536 [Bordley, John Beale (1727-1804)]. *Sketches on Rotation of Crops.* Philadelphia: Charles Cist, 1792. Evans: 24129. Description of English system of crop rotation with diagrams for planting and a design for using a rotation system with American crops.

537 A *Complete Guide for the Management of Bees, through the Year.* Worcester: Isaiah Thomas and Leonard Worcester, 1792. Evans: 24207. Directions for care of bees and hives, with some natural history of bees; also provides directions for manufacturing honey and mead. Illustrations are included.

538 DeWitt, Simeon (1756-1834). *A Map of the State of New York.* Albany: 1792. Evans: 24265 Wheat/Brun: 357. DeWitt was United States geographer-in-chief preceding Thomas Hutchins; he was involved in establishing the boundary between New York and Pennsylvania. According to the DAB his maps of New York "are still accepted as an accurate depiction of the state at the time."

539 Ellicott, Andrew (1754-1820). *A Plan of the City of Washington, in the Territory of Columbia, Ceded by the States of Virginia and Maryland to the United States of America and by Them Established as the Seat of Their Government after the Year 1800.* Philadelphia: 1792. Evans: 24296 Wheat/Brun: 531. Map taken from Ellicott's survey based on the original design of Pierre Charles L'Enfant. This map is usually referred to as the "official" plan of Washington and was reproduced many times.

540 Folie, A.P. *Plan of the Town of Baltimore and Its Environs.* Philadelphia: 1792. Evans: 24323 Wheat/Brun: 521. Map of Baltimore with a key indicating buildings.

541 Hargrove, John. *The Weaver's Draft Book and Clothier's Assistant.* Baltimore: J. Hagerty, 1792. Evans: 24996. Collection of "draughts and receipts" for setting up looms to weave different patterns. It includes recipes for sizing cotton warps and dyeing yarn.

542 Johnson, Gordon. *An Introduction to Arithmetic.* Springfield: Ezra W. Weld, 1792. Evans: 24435 Karpinsky: p. 101. Arithmetic textbook; there were three later editions.

543 New York (State). *The Report of a Committee Appointed to Explore the Western Waters in the State of New York for the Purpose of Prosecuting the Inland Lock Navigation.* Albany: Barber and Southwick, 1792. Evans: 24604. Description of Mohawk River from Schenectady to Fort Schuyler with suggestions for building a canal route and some estimates of cost.

544 Northern Inland Lock Navigation Company. *A Report of the Committee Appointed by the Directors of the Northern Inland Lock Navigation Company in the State of New York to Examine Hudson's River.* [New York]: W. Durell, [1792]. Evans: 24635. Description of the Hudson River from the mouth of a creek at Troy to Lansingburgh; suggestions for making the river navigable with some suggested costs and procedures.

545 Rush, Benjamin (1745-1813). *An Account of the Sugar Maple Tree of the United States.* Philadelphia: R. Aitken and Son, 1792. Evans: 24761. Paper read before the American Philosophical Society; includes methods of tapping trees and manufacturing sugar. Rush argues for the superior quality and price of maple sugar and claims it has medicinal value; he also includes an attack on the use of slave labor in the manufacture of West Indian sugar.

546 Society for the Promotion of Agriculture, Arts and Manufactures. *Transactions of the Society Instituted for the Promotion of Agriculture, Arts, and Manufactures. Part I.* New York: Childs and Swaine, 1792. Evans: 24605. Articles on agriculture including manures, seed, maple sugar, pest control, and planting. Authors include S.L. Mitchill, Robert Livingston, George Logan, Edward Hand, Ezra L'Hommedieu, Jonathen Havens, Simeon DeWitt, and Walter Rutherford.

547 Vinall, John (1736-1823). *The Preceptor's Assistant or Student's Guide, Being a Systematical Treatise of Arithmetic, Both Vulgar and Decimal, Calculated for the Use of Schools, Counting Houses, and Private Families.* Boston: P. Edes, 1792. Evans: 24962 Karpinsky: p. 102. Arithmetic text; emphasis on currency.

548 Williams, Jonathan, Jr. (1750-1815). *Memoir on the Use of the Thermometer in Navigation.* Philadelphia: R. Aitken and Son, 1792. Evans: 25040. Paper presented to the American Philosophical Sociey; concerned with the use of the thermometer in discovering "banks and soundings." Williams includes tables of data, maps and extracts from his ship journals. Williams was Benjamin Franklin's great-nephew and worked with Franklin on some of his later experiments.

549 [Young, Arthur (1741-1820)]. *Rural Economy, or Essays on the Practical Parts of Husbandry, Designed to Explain Several of the Most Important Methods of Conducting Farms of Various Kinds, Including Many Useful Hints to Gentleman Farmers, Relative to the Economical Management of Their Business.* Burlington: Isaac Neale, 1792. Evans: 25061. Collection of agricultural essays. Submitted to the Society for Promoting Agriculture through the influence of George Washington.

1793

550 *The Art of Cheese Making, Taught from Actual Experiments in Which More and Better Cheese May Be Made from the Same Quantity of Milk.* Concord, N.H.: George Hough, 1793. Evans: 25123. Directions for an agricultural audience.

551 Blanchard, Jean Pierre Baptiste (1753-1809). *Journal of My Forty-Fifth Ascension, Being the First Performed in America, on the Ninth of January, 1793.* Philadelphia: Charles Cist, 1793. Evans: 25207. Description of a balloon flight, including preparations plus narrative of ascension.

552 Clendinin, John. *The Practical Surveyor's Assistant, in Two Parts.* Philadelphia: 1793. Evans: 25305 Karpinsky: p. 102. Tables of latitude and departure.

553 *An Easy Method of Working by the Plain and Sliding Rules.* Lansingburgh: Silvester Tiffany, 1793. mp: 26740 Bristol: 8329. Slide rule instructions; described as intended for "architectors."

554 [Freeman, James (1759-1835)]. *Remarks on the American Universal Geography.* Boston: Belknap and Hall, 1793. Evans: 25510. Critique of Morse's *American Geography* (see 509); accuses Morse of inconsistency, inaccuracy, poor judgment, bias, and carelessness and includes a page-by-page listing of errors. Freeman and Morse were involved in a bitter dispute over Unitarianism at the time of publication.

555 *The Golden Cabinet, Being the Laboratory or Handmaid of the Arts.* Philadelphia: William Spotswood, 1793. Evans: 25551. Collection of miscellaneous directions for manufacturing processes including "gilding, silvering, bronzing, japanning, lacquering, and the staining different kinds of substances with all the variety of colors." It also includes directions for drawing and painting as well as working glass and paste.

556 Imlay, Gilbert (c.1754-c.1828). *A Topographical Description of the Western Territory of North America, Containing a Succinct Account of Its Climate, Natural History, Population, Agriculture, Manners and Customs, with an Ample Description of the Several Divisions into Which That Country Is Divided.* New York: Samuel Campbell, 1793. Evans: 25648 Hazen: 5512. Geographical description and maps of the Carolinas, Georgia, and Florida as far west as Spanish Louisiana; includes "laws and government" of Kentucky. Imlay includes not only his own work, but also Filson's history of Kentucky (see 463) and two works by Thomas Hutchins on Virginia, Maryland, North Carolina, Louisiana, and Florida (see 465 and 484).

557 Massachusetts Society for Promoting Agriculture. *Laws and Regulations of the Massachusetts Society for Promoting Agriculture.* Boston: Isaiah Thomas and Ebenezer T. Andrews, 1793. Evans: 25794.

Includes constitution, officers, premiums offered for agricultural discoveries, and extracts from publications of other agricultural societies.

558 Merrill, Phinehas (1767-1814). *Plan of the Town of Strathem.* n.p.: 1793. Not included in Readex Wheat/Brun: 184. Map of Strathem, New Hampshire with houses, roads, and topography.

559 Merrill, Phinehas (1767-1815). *The Scholar's Guide to Arithmetic, Being a Collection of the Most Useful Rules.* Exeter: Henry Ranlet, 1793. Evans: 25806 Karpinsky: p. 103. Arithmetic textbook, including plane and solid geometry; there were six more editions, the last in 1819.

560 Pike, Nicholas (1743-1819). *Abridgement of the New and Complete System of Arithmetick [sic], Composed for the Use, and Adapted to the Commerce of the Citizens of the United States.* Newburyport: J. Mycall, 1793. Evans: 26002 Karpinsky: p. 105. Abridgement of Pike's arithmetic text (see 495); particularly in abridged form it was "the first widely popular treatise by a native American," according to Karpinsky. There were several more editions, the last in 1832.

561 Sarjeant, Thomas. *The Federal Arithmetician, or the Science of Numbers, Improved.* Philadelphia: Thomas Dobson, 1793. Evans: 26137 Karpinsky: p. 108. Arithmetic textbook; includes section on "Federal Money" and "the Exchange," which includes a subsection on "Weights and Measures."

562 [Smith, Daniel (1748-1818)]. *A Short Description of the Tenassee [sic] Government, or the Territory of the United States South of the River Ohio, to Accompany and Explain a Map of That Country.* Philadelphia: Mathew Carey, 1793. Evans: 26168. Geography of Tennessee; sold without a map until Smith's map was published in Carey's *General Atlas* (see 591). Smith was a Tennessee pioneer, secretary of the territory southwest of the Ohio, one of the authors of Tennessee's constitution and Jackson's successor in the United States Senate.

563 Society for Promoting the Manufacture of Sugar from the Sugar Maple Tree. *Constitution of the Society for Promoting the Manufacture of Sugar from the Sugar Maple Tree.* Philadelphia: Robert Aitken, 1793. mp: 46876 Bristol: 8476. Organizational rules of the society.

564 Townsend, David. *Principles and Observations Applied to the Manufacture and Inspection of Pot and Pearl Ashes.* Boston: Isaiah Thomas and Ebenezer T. Andrews, 1793. Evans: 26270. Description of common types of pot and pearl ash as well as the manufacturing process. An appendix is a description of pearl ash and pot ash furnaces.

Townsend is described as "Inspector of Pot and Pearl Ashes for the Commonwealth of Massachusetts."

565 *A Treatise on Silkworms.* New York: T. Allen, 1793. Evans: 26276. Directions for growing silkworms both indoors and outdoors; arranged in question and answer form. It includes the act establishing the New York Society for the Promotion of Agriculture, Arts, and Manufacture.

566 United States Congress. *The Committee to Whom Was Referred the Motions before the Senate Relative to the Weights and Measures--Report.* Philadelphia: John Fenno, 1793. Evans: 26348. Recommendations for establishing uniform weights and measures (see 517). Includes methods for deriving standards.

567 Whitelaw, James (1748-1829). *A Map of the State of Vermont.* [Boston]: 1793. Evans: 26478 Wheat/Brun: 194. Shows counties and their boundaries, but not townships. Also included in Williams' *Natural and Civil History of Vermont* (see 587).

568 Whitney, Peter (1744-1816). *The History of the County of Worcester, in the Commonwealth of Massachusetts.* Worcester: Isaiah Thomas, 1793. Evans: 26481 Wheat/Brun: 211. Geography and map of Worcester County.

569 Williams, John Foster (1743-1814). *Capt. J. F. Williams' Apparatus for Extracting Fresh Water from Salt Water.* Boston: Benjamin Russell, [1793]. mp: 46947 Bristol: 8558. Description of experiments along with a diagram of the apparatus; originally communicated to the Boston Marine Society in 1792.

570 Workman, Benjamin. *Elements of Geography Designed for Young Students in That Science.* 4th ed. Philadelphia: John McCullough, 1793. Evans: 26509 Hazen: 11060. Geography textbook including maps.

1794

571 Barker, Elihu. *A Map of Kentucky from Actual Survey.* Philadelphia: Mathew Carey, 1794. Evans: 26616 Wheat/Brun: 641. Map of Kentucky and part of the Northwest Territory; indicates roads and trails with notes. It was later republished in London.

572 Bordley, John Beale (1727-1804). *Intimations on Manufactures.* Philadelphia: Charles Cist, 1794. Evans: 26681. Discussion of American manufacturing.

573 Carey, Mathew (1760-1839). *A General Atlas for the Present War.* Philadelphia: Mathew Carey, 1794. Evans: 26741. Includes six maps of Europe and a chart of the West Indies.

574 Colles, Christopher (1738-1821). *The Geographical Ledger and Systematized Atlas.* New York: John Buel, 1794. Evans: 26781. Wheat and Brun call it "a large, indexed topographical map intended to cover the United States and possibly more." Letters on the maps referred to sections of the book; the map sections themselves were supplied separately. The copy reproduced by Readex includes only New York.

575 Davies, Benjamin (fl. 1774-1806). *Some Account of the City of Philadelphia, the Capital of Pennsylvania and Seat of the Federal Government.* Philadelphia: Richard Folwell, 1794. Evans: 26853. Geography and history of Philadelphia; includes maps.

576 DeWitt, Simeon. *A Plan of the City of Albany, Surveyed at the Request of the Mayor, Aldermen, and Commonality.* Albany: 1794. Not included in Readex Wheat/Brun: 359. Wheat and Brun call it "a detailed plan of the city"; it includes a key identifying several locations.

577 Doolittle, Amos (1754-1820). *Connecticut from the Best Authorities.* New Haven: A. Doolittle, [1794]. Not included in Readex Wheat/Brun: 274. Includes the area in dispute between Connecticut and New York.

578 Ellicott, Andrew (1754-1820). *Territory of Columbia.* [Philadelphia: Joseph T. Scott, 1794]. Not included in Readex Wheat/Brun: 535. First map of the District.

579 [Lear, Tobias (1762-1816)]. *Observations on the River Potomack, the Country Adjacent, and the City of Washington.* New York: Loudon and Brown, 1794. Evans: 27209. Geography of the District of Columbia; Lear was Washington's private secretary and the *Observations* are "probably the earliest separate monograph on the District of Columbia " (DAB). The work has sometimes been attributed to Andrew Ellicott (see, for example, Meisel: 3: p. 358).

580 *Mathematical Tables: Difference of Latitude and Departure, Logarithms from 1 to 10,000.* Philadelphia: Joseph Crukshank, 1794. Evans: 27301. Tables without any accompanying explanation.

581 Patterson, Robert (1743-1824). *A New Table of Latitude and Departure, for Every Degree and Five Minutes of the Quadrant and for Every Point and Quarter of the Compass.* Philadelphia: Stewart and Cochran, 1794. mp: 47152 Bristol: 8785 Karpinsky: p. 109. Table followed by directions for use; Patterson was professor of mathematics at the University of Pennsylvania.

582 Philadelphia Society for Promoting Agriculture. *Outlines of a Plan for Establishing a State Society of Agriculture in Pennsylvania.* Philadelphia: Charles Cist, 1794. Evans: 27512. Proposal for founding a state agricultural society. The Philadelphia Society was founded in 1785 as an offshoot of the American Philosophical Society; its object was to increase and promote agricultural production. Similar organizations were founded in New Jersey, South Carolina, and New York.

583 *A Short and Easy Guide to Arithmetic, Particularly Adapted to the Use of Farmers and Tradesmen in the United States of America.* Boston: Samuel Hall, 1794. Evans: 27700 Karpinsky: p. 108. Arithmetic textbook, including some sample forms for wills, deeds, promissory notes, etc.

584 **Society for the Promotion of Agriculture, Arts, and Manufactures.** *Transactions of the Society Instituted in the State of New York for the Promotion of Agriculture, Arts, and Manufactures. Part II.* New York: Childs and Swaine, 1794. Evans: 27400. Articles on fishing, crops, cultivation, animal husbandry, geography, farming implements, and measurement; authors include S.L. Mitchilll, Robert Livingston, Ezra L'Hommedieu, Peter Delabigarre, John Smith, John W. Watkins, John Stevens, Henry Muhlenberg, and William Denning.

585 **Truxtun, Thomas (1755-1822).** *Remarks, Instructions, and Examples Relating to the Latitude and Longitude; Also the Variation of the Compass.* Philadelphia: Thomas Dobson, 1794. Evans: 27823. Navigation manual with maps; includes favorable routes, winds and currents, and appendices on masting of warships and duties of naval officers. Truxton was a prominent naval officer.

586 **Varle, Peter.** *To the Citizens of Philadelphia, This Plan of the City and Its Environs Is Respectfully Dedicated by the Editor.* [Philadelphia: c. 1794]. Not included in Readex Wheat/Brun: 465. Map of Philadelphia including the names of some householders beside their houses. The city is shown as it was circa 1794.

587 **Williams, Samuel (1743-1817).** *The Natural and Civil History of Vermont.* Walpole: Isaiah Thomas and David Carlisle, Jr., 1794. Evans: 28094 Meisel: 3: p. 358. Geography of Vermont, including climate, plants, and animals.

1795
588 **[Adgate, Andrew (d.1793)].** *A Lecture Containing a Short History of Mechanics and of Useful Arts and Manufactures.* Catskill: 1795. Evans: 28154. Delivered to the Mechanics of Philadelphia and published under the pseudonym "Absalom Aimwell." It is a semi-humorous description of various mechanical specialties in ancient and biblical texts.

589 **Anderson, Alexander (1775-1870).** *Map of the State of Kentucky with the Adjoining Territories.* New York: Smith, Reid, and Wayland, 1795. Evans: 28191 Wheat/Brun: 646. Copied from a map

by John Russell published in London. Wheat and Brun call it "the best map produced of the area so far."

590 [Bingham, Caleb. (1757-1817)]. *An Astronomical and Geographical Catechism.* Boston: S. Hall, 1795. Evans: 28303. Textbook for children arranged in question and answer format; based on Morse's *Geography Made Easy* (see 468).

591 Carey, Mathew (1760-1839). *The General Atlas for Carey's Edition of Guthrie's Geography, Improved.* Philadelphia: Mathew Carey, 1795. mp: 47370 Bristol: 9043 Not included in Readex. Collection of maps: world, European, Asian, African, and North American. Includes the first map of Tennessee, drawn by Daniel Smith.

592 *Carey's American Atlas.* Philadelphia: Mathew Carey, 1795. Evans: 28390. Maps of individual states, South America, Northwest Territories, and chart of the West Indies.

593 Carleton, Osgood (1741-1816). *An Accurate Map of the Commonwealth of Massachusetts, Exclusive of the District of Maine, Compiled Pursuant to an Act of the General Court from Actual Surveys of the Several Towns, &c., Taken by Their Order.* Boston: O. Carleton and I. Norman, [1795]. Not included in Readex Wheat/Brun: 214. Includes counties, townships, roads and rivers.

594 Carleton, Osgood (1741-1816). *An Accurate Map of the District of Maine, Being Part of the Commonwealth of Massachusetts.* Boston: Thomas and Andrews, 1795. Evans: 28391 Wheat/Brun: 170. Shows towns, roads, rivers, some buildings, and other topographical features.

595 Chaplin, Joseph. *The Trader's Best Companion.* Newburyport, MA: William Barrett, 1795. Evans: 28405 Karpinsky: p. 110. Application of rules of arithmetic to federal currency; includes tables for computing prices.

596 Dwight, Nathaniel (1770-1831). *A Short but Comprehensive System of the Geography of the World by Way of Question and Answer.* Hartford: Elisha Babcock, 1795. Evans: 28606. Geography textbook in question and answer format; published while Dwight was a physician. It was "enthusiastically received" (DAB) and valued as a teaching text.

597 Evans, Oliver (1755-1819). *The Young Mill-Wright and Miller's Guide.* Philadelphia: 1795. Evans: 28644. Instructions for building and operating a mill; includes principles of mechanics and hydraulics for water-mills, with illustrations. Evans built the first successful steam engine for use in a mill.

598 Griffith, Dennis. *Map of the State of Maryland.* Philadelphia: J.
Vallance, 1795. Evans: 28772 Wheat/Brun: 511. Detailed map; Wheat
and Brun call it "the best of Maryland to this time."

599 Harris, Caleb. *A Map of the State of Rhode Island, Taken Mostly
from Surveys.* Providence: Carter and Wilkinson, 1795. Evans: 28803
Wheat/Brun: 251. According to Wheat and Brun "a fine political map of
the state showing also a few topographical features."

600 Lewis, Samuel (fl. 1796). *A Map of the United States,
Compiled Chiefly from the State Maps and Other Authentic Information.*
1795. Evans: 32378 Wheat/Brun: 123. Includes land sold by New
York to Pennsylvania, displayed separately from the two states.

601 Lewis, Samuel (fl. 1796). *The State of New York.*
[Philadelphia: M. Carey, 1795]. mp: 47485 Bristol: 9171 Wheat/Brun:
362. Western boundary based on 1786 agreement with Pennsylvania;
the same map was included in *Carey's American Atlas* (see 592).

602 *A Map of the Military Lands and 20 Townships in the Western Part of
the State of New York.* n.p.: [c. 1795]. Not included in Readex
Wheat/Brun: 365. Includes major land purchases and grants as well as
major settlements.

603 Maxcy, Jonathan (1768-1820). *An Oration Delivered before the
Providence Association of Mechanics and Manufacturers, at the Annual
Election, April 13, 1795.* Providence: Bennett Wheeler, 1795. Evans:
29053. Speech on need for and dignity of the "mechanic arts"; Maxcy
was president of Rhode Island College (Brown University) and a
prominent orator.

604 Morse, Jedidiah (1761-1826). *Elements of Geography.*
Boston: I. Thomas and E.T. Andrews, 1795. Evans: 29112 Hazen:
7450. Geography text for children; includes a map of the United States
and a world chart.

605 Morse, William. *Mechanical Arts in Thirty-Two Receipts.*
Hartford: Elisha Babcock, 1795. mp: 47494 Bristol: 9183. Directions
for a variety of manufacturing processes, including making varnish,
hardening leather, and dyeing various substances.

606 *Pocket Gazetteer, or a Description of All the Principal Cities and
Towns in the United States of America.* Walpole: Isaiah Thomas and
David Carlisle, 1795. Evans: 29331. Alphabetical listing of cities and
towns of the United States, giving locations.

607 Porter, Augustus (1769-1864). *A Map of Messrs. Gorhan and
Phelps's Purchase Now in the County of Ontario in the State of New
York.* [New Haven?: 1795]. Evans: 29341 Wheat/Brun: 361. Shows

the survey of the purchase into ranges and townships; includes several topographical features and some buildings.

608 Price, Jonathan. *A Description of Occacock Inlet.* Newbern: Francoise-X. Martin, 1795. Evans: 29351 Wheat/Brun: 582. Ocracoke Inlet, showing channels through the shoals.

609 Root, Erastus (1773-1846). *An Introduction to Arithmetic for the Use of Common Schools.* Norwich: Thomas Hubbard, 1795. Evans: 29430 Karpinsky: p. 111. Popular arithmetic textbook; it was republished many times, last in 1814.

610 Schuylkill and Susquehanna Canal Navigation Company. *An Historical Account of the Rise, Progress and Present State of the Canal Navigation in Pennsylvania.* Philadelphia: Zachariah Poulson, Jr., 1795. Evans: 29474. Description of canals with mileage, maps, and various official communications.

611 Scott, Joseph. *The United States Gazetteer, Containing an Authentic Description of the Several States.* Philadelphia: F. and R. Bailey, 1795. Evans: 29476 Hazen: 9368 Meisel: 3: p. 359. Geographical dictionary with maps.

612 Smith, Charles (1768-1808). *Universal Geography Made Easy. or a New Geographical Pocket Companion.* New York: Wayland and Davis, 1795. Evans: 29521. Directory of terms, with maps and brief descriptions.

613 Sterry, Consider (1761-1817) and John (1766-1823). *A Complete Exercise Book in Arithmetic.* Norwich: John Sterry and Co., 1795. Evans: 29566 Karpinsky: p. 113. Arithmetic textbook.

614 *The Stranger's Assistant and the School-Boy's Instructor.* New York: George Forman, 1795. Evans: 29576 Karpinsky: p. 110. Arithmetic textbook.

615 Sullivan, James (1744-1808). *The History of the District of Maine.* Boston: I. Thomas and E.T. Andrews, 1795. Evans: 29589. Geography of Maine with maps; Sullivan was later governor of Massachusetts and one of the founders of the Massachusetts Historical Society.

616 [Tucker, St. George (1752-1827)]. *A Letter to the Rev. Jedidiah Morse, A.M., Author of the American Universal Geography.* Richmond: Thomas Nicolson, 1795. Evans: 29662. Attack upon Morse's *American Universal Geography* (see 641); Tucker charges Morse with plagiarizing Jefferson and with inaccuracy regarding Virginia.

617 Van Der Kemp, Francis Adrian (1752-1829). *Speech of Fr. Adr. Van der Kemp at a Meeting, the First of June, One Thousand, Seven Hundred and Ninety-Five, at Whitestown, for the Institution of a Society of Agriculture.* Whitestown: Oliver P. Easton, 1795. Proposal for a society of agriculture and natural history.

618 [Webster, Noah (1758-1843)]. *The Farmer's Catechism.* Canaan: Elihu Phinney, 1795. Evans: 29855. "Rules of Husbandry" for use in schools; only the title page is included in Readex (see 698, *The Little Reader's Assistant,* for a complete version).

619 Wilkinson, William (1760-1852). *The Federal Calculator and American Ready Reckoner.* Providence: Carter and Wilkinson, 1795. Evans: 29892 Karpinsky: p. 113. Description of decimal currency: value of cents in state currencies, value of gold, tables of interest, values in dollars and cents of various quantities.

1796

620 *The American Atlas.* New York: John Reid, 1796. Evans: 31078. Includes maps of states, North and South America, the United States, and the West Indies.

621 Blanchard, Jean Pierre Baptiste (1753-1809). *The Principles, History, and Use of Air-Balloons.* New York: C.C. Van Alen, 1796. Evans: 30099. History and principles of balloon flight, along with descriptions of construction and fuel; ends with a prospectus for a balloon flight from New York.

622 [Bordley, John Beale (1727-1804)]. *Sketches on the Rotation of Crops and Other Rural Matters.* Philadelphia: Charles Cist, 1796. Evans: 30103. Description of various crop rotation plans; also includes other miscellaneous discussions of agricultural topics including manure, cattle stalls, and ice-houses.

623 Bradley, Abraham, Jr (d. 1838). *Map of the United States Exhibiting Post Roads and Distances by Abraham Bradley Jr., the First Sheet Comprehending the Nine Northern States with Parts of Virginia and the Territory North of Ohio.* Philadelphia: 1796. Evans: 30123 Wheat/Brun: 127. Wheat and Brun call it "the best map of the new nation to appear up to this time." Bradley was associated with the Post Office Department; his maps were official postal maps.

624 Burrill, George Rawson (1770-1818). *An Oration Delivered before the Providence Association of Mechanics and Manufacturers at Their Annual Election, April 11, 1796.* Providence: Bennett Wheeler, 1796. Evans: 30145. Celebration of "mechanic arts" and scientific principles.

625 Carey, Mathew (1760-1839). *Carey's American Pocket Atlas.* Philadelphia: Mathew Carey, 1796. Evans: 30161. Maps and descriptions of seventeen states, the Northwest Territory, and the country as a whole.

626 Carey, Mathew (1760-1839). *Carey's General Atlas.* Philadelphia: Mathew Carey, 1796. Evans: 30162. Forty-five maps, including Europe, Asia, Africa, the United States, individual states, the North Pole, and "Captain Cook's Discoveries."

627 Carleton, Osgood (1742-1816). *Plan of the Town of Boston from Actual Survey.* Boston: Manning and Loring, 1796. Evans: 30164 Wheat/Brun: 246. Drawn in 1795 after a new survey by Carleton; first published in the *Boston Directory* of 1796.

628 [Condy, Jonathan Williams (1770-1828)]. *Description of the River Susquehanna with Observations on the Present State of Its Trade and Navigation, and Their Practicable and Probable Improvement.* Philadelphia: Zachariah Poulson, Jr., 1796. Evans: 30338. Geography of the Susquehanna and its environs; includes maps.

629 [Dabney, John, comp.] *An Address to Farmers.* Newburyport: Blunt and March, 1796. Evans: 30307. Collection of essays on various agricultural subjects including manure, labor, nurseries, orchards, cider, and the "Character of a Complete Farmer."

630 Denison, J. *Map of the States of Maryland and Delaware.* Boston: 1796. Evans: 30334 Not included in Readex Wheat/Brun: 514. Includes some roads, county boundaries, and settlements.

631 Ellicott, Andrew (1754-1820). *Several Methods by Which Meridional Lines May Be Found with Ease and Accuracy.* Philadelphia: Thomas Dobson, 1796. Evans: 30385. Textbook for surveyors.

632 Fairman, Gideon. *A Map of the Oneida Reservation, Including the Lands Leased to Peter Smith.* Albany: [1796?]. W/B: 369. Shows surveys of the eastern and southern parts of the reservation, including the new Stockbridge Township.

633 Furlong, Lawrence (1734-1806). *The American Coast Pilot.* Newburyport: Blunt and March, 1796. Evans: 30464. Sailing directions from Boston along the Atlantic coast; includes tide tables.

634 Lewis, Samuel (fl. 1796). *A Map of Part of the Northwest Territory of the United States, Compiled from Actual Survey and the Best Information.* Philadelphia: Mathew Carey, 1796. Evans: 30691 Wheat/Brun: 678. Map of the territory as far west as Lake Michigan, north to Canada and south to Kentucky; includes seven ranges and lands ceded by Indian tribes.

635 Massachusetts Society for Promoting Agriculture. *Boston, Nov. 1, 1796.* [Boston: 1796]. Evans: 30772. Letter to clergy recommending a new series of pamphlets from the society.

636 Massachusetts Society for Promoting Agriculture. *Rules and Regulations of the Massachusets Society for Promoting Agriculture.* Boston: Thomas Fleet, 1796. Evans: 30773. Rules and names of officers; also a collection of "Communications" to the society on subjects such as livestock, dairy farming, and apple-growing.

637 Molineux, Frederic. *Plan of the Town of Erie.* Philadelphia: 1796. Evans: 30805 Wheat/Brun: 449. Map of Erie, Pennsylvania indicating some streets and numbered lots.

638 Moore, Samuel (c.1737-1810). *An Accurate System of Surveying.* Litchfield: T. Collier, 1796. Evans: 30810 Karpinsky: p. 114. Surveying textbook.

639 Moreau de Saint-Mery, Mederic Louis Elie (1750-1819). *A Topographical and Political Description of the Spanish Part of Saint-Domingo. In two volumes.* Philadelphia: 1796. Evans: 30818. Geography of Santo Domingo, with maps; in French with side-by-side translation by William Cobbett. Moreau de Saint-Mery was a French emigre, a former resident of Santo Domingo who was forced to flee during the slave rebellion.

640 Morse, Jedidiah (1761-1826). *The American Universal Geography, or a View of the Present State of All the Empires, Kingdoms, States and Republics in the Known World and of the United States of America in Particular. Part 1.* Boston: Isaiah Thomas and Ebenezer T. Andrews, 1796. Evans: 30824 Hazen: 7449. Second edition of Morse's *American Geography* (see 509); the section on natural history was improved with additions from Francisco Clavigero, S.L. Mitchill, Manasseh Cutler, and Joshua Fisher. The earlier maps were retained, but some new ones were added.

641 Morse, Jedidiah (1761-1826). *The American Universal Geography, or a View of the Present State of All the Empires, Kingdoms, States and Republics in the Known World and of the United States of America in Particular. Part 2.* Boston: Isaiah Thomas and Ebenezer T. Andrews, 1796. Evans: 30823. See 640.

642 Pattillo, Henry (1726-1801). *A Geographical Catechism.* Halifax, NC: Abraham Hodge, 1796. Evans: 30963. Geography text with question and answer format.

643 *Plat of the Seven Ranges of Townships, Being Part of the Territory of the United States N.W. of the River Ohio, Which by a Late Act of*

Congress Are Directed to Be Sold. [Philadelphia: Mathew Carey, 1796]. Evans: 30918 Wheat/Brun: 676. Map of the survey dividing public lands into ranges and townships for the first time according to the Ordinance of 1785. Thomas Hutchins conducted the surveys of the first four and part of the fifth range.

644 **Ruddock, Samuel Abial.** *A Geographical View of All the Post Towns in the United States of America and Their Distances from Each Other, According to the Establishment of the Postmaster General.* [Boston:1796]. Evans: 31132. Table of mileage.

645 *Rules of Arithmetic Containing All the Useful Tables in Money, Weights and Measures.* Philadelphia: D. Humphreys, 1796. Evans: 31139 Karpinsky: p. 114. Arithmetic reference book containing tables of measurement.

646 **Scott, Joseph.** *An Atlas of the United States.* Philadelphia: Francis and Robert Bailey, 1796. mp: 47908 Bristol: 9695. United States and state maps, also Northwest and Southwest Territories and Kentucky.

647 **Temple, Samuel (1770-1816).** *A Concise Introduction to Practical Arithmetic.* Boston: Samuel Hall, 1796. Evans: 31276 Karpinsky: p. 115. Arithmetic text with emphasis on federal currency; there were several more editions, the last in 1827.

648 **United States Congress.** *Papers Relative to an Application to Congress for an Exclusive Right of Searching for and Working Mines in the North-West and South-West Territory by N[icholas} L. Roosevelt and J[acob] Mark.* [Philadelphia]: Samuel Harrison Smith, [1796]. Evans: 31473 Hazen: 10509. Includes "Memorial" by Roosevelt and Mark asking to be granted rights of mining and smelting ore; also supporting letter by William Langworthy discussing financial risks, laws of other countries regarding mining, and some description of possible ores. The favorable Congressional Committee report is included.

649 **United States Congress.** *Report of the Committee to Whom Were Referred So Much of the Report of the Secretary of State Made the 13th of July, 1790, and the Message from the President of the United States on the 8th of January, 1795, As Related to Weights and Measures.* [Philadelphia: 1796]. Evans: 31390. Proposal for establishing uniform weights and measures; includes proposal for a metric system.

650 **Western and Northern Inland Lock Navigation Companies.** *Report of the Directors of the Western and Northern Inland Lock Navigation Companies in the State of New York to the*

Legislature. New York: George Forman, 1796. Evans: 31623.
Description of progress on building canal.

651 Whitelaw, James (1748-1829). *A Correct Map of the State of Vermont from Actual Survey.* [Rutland]: 1796. Evans: 31626 Wheat/Brun: 200. Issued as official map of state; republished several times in nineteenth century, as late as 1851.

652 Wright, Benjamin. *A Map of the North East Part of the Town of Mexico, the Property of George Scriba, Situated on the Lake Ontario, About 13 Miles East from Oswego and 20 from Rotterdam, Oneida Lake.* n.p.: P.R. Maverick, 1796. Not included in Readex Wheat/Brun: 370. Plan of Mexico, New York showing streets and many blocks divided into lots

1797

653 *The American Farmer's Guide, or a New and Excellent Treatise on Agriculture.* Philadelphia: Jacob Johnson, 1797. Evans: 31718. Reference work with directions for cultivating several crops.

654 Barker, W. *Plan of Beula, Pennsylvania.* n.p.: [c. 1797]. Not included in Readex Wheat/Brun: 452. Includes a short description of the town; some streets are named.

655 Benjamin, Asher (1773-1845). *The Country Builder's Assistant, Containing a Collection of New Designs of Carpentry and Architecture Which Will Be Particularly Useful to Country Workmen in General.* Greenfield, MA: Thomas Dickman, 1797. Evans: 31797. Series of plates of architectural details with directions for drawing and building them. Benjamin's influence was widespread and he helped to popularize the "late colonial" style in New England (DAB).

656 Birch, William. *A Plan of the Island of Burlington and a View of the River Delaware.* Philadelphia: W. Birch, 1797. Not included in Readex Wheat/Brun: 419. Map of Burlington, New Jersey indicating many street names and town buildings; Birch is more famous as a miniaturist than as a mapmaker.

657 [Bordley, John Beale (1727-1804)]. *Intending to Retire. . . .* [Philadelphia: 1797]. mp: 48064 Bristol: 9869. Discussion of ideal design and organization of a farm for someone with little agricultural knowledge and limited land.

658 [Bordley, John Beale (1727-1804)]. *Queries Selected from a Paper of the Board of Agriculture in London, on the Nature and Principles of Vegetation, with Answers and Observations by J.B.B.* [Philadelphia: Charles Cist, 1797.] Evans: 31845. Collection of agricultural discussions in question and answer format.

659 Carleton, Osgood (1742-1816). *An Accurate Plan of the Town of Boston and Its Vicinity, Exhibiting a Ground Plan of All the Streets, Lanes, Alleys, Wharves, and Public Buildings in Boston, from the Actual Surveys of the Publisher, Also Boston from the Actual Surveys of the Publisher, Also Part of Charlestown and Cambridge from the Surveys of Samuel Thompson, Esqr. and Part of Roxbury and Dorchester from Those of Mr. Witherington [sic].* Boston: O. Carleton, 1797. Not included in Readex Wheat/Brun: 248. Based on Carleton's 1795 survey, the first since Bonner's map (see 383); the plate was later cut down and the map reissued in 1800 under another title. The surveyor's name was actually Matthew Withington.

660 Galland, John (fl. 1796-1817). *[Plan of Baltimore.* Baltimore: George Keatinge, 1797].* mp: 48132 Bristol: 10073. Not included in Readex Wheat/Brun: 524. Said to include streets and public buildings. No copy is known.

661 Henfrey, Benjamin. *A Plan with Proposals for Forming a Company to Work Mines in the United States, and to Smelt and Refine the Ores, Whether of Copper, Lead, Tin, Silver, or Gold.* Philadelphia: Snowden and McCorkle, 1797. Evans: 32245 Hazen: 4984 Meisel: 3: p.360. Attempt to interest investors in a mining company; some description of mineral deposits is included.

662 Hills, John. *This Plan of the City of Philadelphia and Its Environs (Shewing the Improved Parts) Is Dedicated to the Mayor, Aldermen, and Citizens Thereof.* [Philadelphia: 1797]. Evans: 32253 Wheat/Brun: 468. Wheat and Brun describe it as "a fine detailed, topographical map"; includes wharves and their boundaries.

663 Kendal or Kendall, David (1768-1853). *The Young Lady's Arithmetic.* Leominster, MA: Charles Prentiss, 1797. Evans: 32333 Karpinsky: p. 118. Arithmetic textbook.

664 Lee, Chauncey (1763-1842). *The American Accomptant [sic], Being a Plain, Practical and Systematic Compendium of Federal Arithmetic.* Lansingburgh: William W. Wands, 1797. Evans: 32366 Karpinsky: p. 118. Arithmetic text; first work to use the dollar sign in print.

665 Logan, George (1753-1821). *Fourteen Agricultural Experiments to Ascertain the Best Rotation of Crops.* Philadelphia: Francis and Robert Bailey, 1797. Evans: 32382. Description of the results of seven years of crop rotations; addressed to the Philadelphia Society for Promotion of Agriculture. Logan was a gentleman farmer, member of the American Philosophical Society, and later senator from Pennsylvania.

666 Morse, Jedidiah (1761-1826). *The American Gazetteer.*
Boston: Samuel Hall, 1797. Evans: 32509 Hazen: 7451. Geography
of the United States in encyclopedia form; includes map. It was
republished several times.

667 Morse, Jedidiah (1761-1826). *A Description of the Soil,
Productions, Commercial, Agricultural and Local Advantages of the
Georgia Western Territory.* Boston: Thomas and Andrews, 1797.
Evans: 32510 Hazen: 7452. Material taken from *The American
Gazeteer* (see 666).

668 Nichols, Francis (fl. 1797-1820). *A Treatise of Practical
Arithmetic and Book Keeping.* Boston: Manning and Loring, 1797.
Evans: 32583 Karpinsky: p. 119. Arithmetic textbook.

669 Peale, Charles Willson (1741-1828). *An Essay on Building
Wooden Bridges.* Philadelphia: Francis Bailey, 1797. Evans: 32642.
Directions for building simple bridges; illustrated.

670 Peters, Richard (1744-1828). *Agricultural Enquiries on
Plaister [sic] of Paris.* Philadelphia: Charles Cist, 1797. Evans: 32673
Hazen: 8247. Discussion of gypsum as fertilizer; Peters was a Judge of
the United States District Court in Pennsylvania and the first president
of the Philadelphia Society for Promotion of Agriculture. The
Agricultural Enquiries "exercized wide influence in introducing the
culture of clover and other grasses" (DAB).

671 Pinkham, Paul. *A Chart of George's Bank, Including Cape Cod,
Nantucket and the Shoals Lying on Their Coast, with Directions for
Sailing over the Same &c.* [New Haven]: Edmund M. Blunt, 1797.
Evans: 32693 Not included in Readex Wheat/Brun: 220. Chart with
navigational notes.

672 [Smith, Charles (1768-1808)]. *The American Gazetteer or
Geographical Companion.* New York: Alexander Menut, 1797. Evans:
32841. Geographical encyclopedia; includes one map.

673 Taylor, Benjamin. *A New and Accurate Plan of the City of New
York in the State of New York in North America.* [New York]: 1797.
Not included in Readex Wheat/Brun: 394. Map of New York as it
appeared in 1796; includes street names, some wards, residences, and
public buildings. Wheat and Brun call it "an excellent topographical
map."

**674 [Todd, John and Zachariah Jess, William Waring, and
Jeremiah Paul].** *The American Tutor's Assistant, or a Compendious
System of Practical Arithmetic.* 3rd ed. Philadelphia: Zachariah
Poulson, Jr., 1797. Evans: 31732 Karpinsky: p. 97. Arithmetic text.
No copy of the first edition is known to exist, but Poulson took out a

copyright for it in 1791. Many editions were printed in the early
nineteenth century, the last in 1830.

675 **United States Congress.** *Report of the Committee to Whom Was
Referred So Much of the Speech of the President of the United States to
Both Houses of Congress As Relates to the Promotion of Agriculture.*
[Philadelphia: 1797]. Evans: 33031. Proposal for an American Society
of Agriculture to stimulate agricultural production.

1798

676 **[Bordley, John Beale (1727-1804)].** *Cattle Pastured and
Soiled in Summer.* [Philadelphia:1798]. mp: 48372 Bristol: 10242.
Comparison of benefits from putting cattle out to pasture or from
keeping them tied up and fed with cut grass ("soiled").

677 **Bordley, John Beale (1727-1804).** *Country Habitations.*
[Philadelphia: 1798]. Evans: 33435. Discussion of principles of
construction for farmhouses; includes one diagram.

678 **Carleton, Osgood (1742-1816).** *An Accurate Plan of 189,126
Acres of Land on Penobscot River, Being the Purchase from the
Penobscot Indians by Government on Each Side Said River, Together
with Two Gores of Land, One on Each Side.* n.p.: c. 1798. Not
included in Readex Wheat/Brun: 177. Map of tract of land northeast of
contemporary Boston.

679 **Conewago Canal Company.** *An Account of the Conewago
Canal, on the River Susquehanna.* Philadelphia: William Young, 1798.
Evans: 33551. History of canal, along with geography and construction
costs; includes act of incorporation.

680 **Dearborn, Benjamin (1754-1838).** *A Description of the
Vibrating Steelyard or Just Balance.* [New Bedford: 1798]. Evans:
35388. Description of Dearborn's invention, including directions for
use and comparison with conventional "steelyard and scales."

681 **DeWitt, Benjamin (1774-1819).** *A Memoir on the Onondaga
Salt Springs and Salt Manufactories in the Western Part of the State of
New York.* Albany: Loring Andrews, 1798. Evans: 33632 Hazen:
3166. Description of salt manufacturing.

682 **Fennell, James (1766-1816).** *Description of the Principles and
Plan of Proposed Establishments of Salt Works, for the Purpose of
Supplying the United States with Home Made Salt.* Philadelphia: J.
Bioren, 1798. Evans: 33729. Proposal for building a salt works,
including probable payback time for investors. Fennell was a prominent
actor who patented a device for extracting salt from sea water; however,
his enterprise was unsuccessful.

683 Gibson, James (fl. 1798). *Atlas Minimus, or a New Set of Pocket Maps of Various Empires, Kingdoms, and States with Geographical Extracts Relative to Each.* Philadelphia: Mathew Carey, 1798. Evans: 33794. Collection of maps with explanations.

684 Gilpin, George. *Plan of the Town of Alexandria in the District of Columbia.* New York: John V. Thomas, 1798. Not included in Readex Wheat/Brun: 539. Includes some topographical details and street names.

685 Jefferson, Thomas (1743-1826). *A Supplementary Note on the Mould Board, Described in a Letter to Sir John Sinclair of March 23, 1798, Inserted in the American Philosophical Transactions, Vol. 4, and in Maese's Domestic Encyclopaedia Voce Plough.* [Philadelphia: 1798]. Evans: 33934. Description of a plough ("mould board") design and suggestions for manufacturing it. According to Wilson, "Plows embodying principles of Jefferson's formula were the ones turned out in quantity following Jethro Wood's development of a plow with cast-iron replaceable parts." The design led to labor saving for both people and animals, and, ultimately, to increased production. Jefferson was awarded a gold medal from the Societe d'Agriculture du Departement de Seine et Oise for this invention. Wilson describes *Supplementary Note* as "perhaps the first scientific paper of its type produced in the United States."

686 Lambert, John (fl. 1798). *A Short and Practical Essay on Farming.* [Philadelphia: 1798]. Evans: 33974. Primary focus on renewing exhausted land; also planting instructions.

687 Massachusetts Society for Promoting Agriculture. *On the Culture of Potatoes.* Boston: Young and Minns, 1798. Evans: 34084. Extracts from Communications to the British Board of Agriculture.

688 Morse, Jedidiah (1761-1826). *An Abridgement of the American Gazetteer.* Boston: Thomas and Andrews, 1798. Evans: 34143. See 666.

689 New York (State) Society for the Promotion of Agriculture, Arts, and Manufactures. *Transactions. Part III.* Albany: Loring Andrews, 1798. Evans: 34221. Articles on crops, animal husbandry, and natural history. Contributors include Jonathan Havens, Ezra L'Hommedieu, Samuel L. Mitchill, Simeon DeWitt, Noah Webster, and Gardner Baker.

690 Payne, John (fl. 1800). *A New and Complete System of Universal Geography. Additions, Corrections and Improvements by James Hardie. Vol. 1.* New York: John Low, 1798. Evans: 34316 Hazen: 8170. Geography text.

691 Pease, Seth (1764-1819). *A Map of the Connecticut Western Reserve from Actual Survey.* New Haven: Amos Doolittle, 1798. Evans: 34317 Wheat/Brun: 680. Map of country surveyed east of the Cuyahoga River; includes several townships. The map was republished twice more with additional townships.

692 Philadelphia. *Report of the Joint Committee of the Select and Common Councils on the Subject of Bringing Water to the City.* Philadelphia: Zachariah Poulson, 1798. Evans: 34350. A report recommending a Delaware and Schuylkill canal.

693 *Plat of That Tract of Country in the Territory Northwest of the Ohio Appropriated for Military Services and Described in the Act of Congress Intituled [sic] "An Act Regulating the Grants of Land Appropriated for Military Services, and for Propagating the Gospel among the Heathen," Survey'd under the Direction of Rufus Putnam, Surveyor General to the United States.* Philadelphia: [c. 1798]. Not included in Readex Wheat/Brun: 681.   Map of the United States Military District; includes 20 ranges of townships and some Indian towns.

694 Price, Jonathan and John Strother. *To Navigators This Chart, Being an Actual Survey of the Sea Coast and Inland Navigation from Cape Henry to Cape Roman, Is Most Respectfully Inscribed by Price and Strother.* New Bern, NC: 1798. Evans: 31046 Not included in Readex Wheat/Brun: 590. Chart of the coastline with some navigation information regarding Albemarle Sound; includes Wilmington, New Bern, Washington, and Edenten.

695 Rush, Benjamin (1745-1813). *Essays Literary, Moral, and Philosophical.* Philadelphia: Thomas and Samuel Bradford, 1798. Evans: 34495. Miscellaneous essays; includes a history of Pennsylvania with a discussion of its agriculture, as well as Rush's account of the sugar maple (see 545).

696 South Carolina Society for Promoting and Improving Agriculture and Other Rural Concerns. *Address and Rules of the South Carolina Society for Promoting and Improving Agriculture and Other Rural Concerns.* Charleston: Freneau and Paine, 1798. Evans: 34577. "Address" urges planters to experiment with agricultural methods to improve American agriculture; includes Society's rules.

697 Tharp, Peter. *A New and Complete System of Federal Arithmetic.* Newburgh: D. Deniston, 1798. Evans: 34644 Karpinsky: p. 122. Arithmetic textbook.

698 Webster, Noah (1758-1843). *The Little Reader's Assistant.* Northhampton: William Butler, 1798. Evans: 34983. Omnibus textbook for children; Part V is the Farmer's Catechism (see 618).

699 Western Inland Lock Navigation Company. *Report of the Directors of the Western Inland Lock Navigation Company to the Legislature.* Albany: Charles R. and George Webster, 1798. Evans: 35012. Description of progress in canal construction.

700 Williams, W. *First Principles of Geography Selected from Approved Authors.* Charleston: T.B. Bowen, 1798. Evans: 35032. Geography textbook.

701 [Williamson, Charles]. *Description of the Genesee Country.* n.p.: 1798. Evans: 35033. Promotional pamphlet; Williamson was a land speculator and promoter.

1799
702 Bordley, John Beale (1727-1804). *Hemp.* [Philadelphia: 1799]. Evans: 35217. Instructions for growing and processing hemp.

703 Carleton, Osgood (1742-1816). *A New Chart of the Northwest Coast of America. . . .* Boston: W. Norman, [c.1799]. Not included in Readex Wheat/Brun: 64. Chart of the Pacific coast from Cook's inlet in Alaska south to Cape St. Lucas in lower California; includes soundings.

704 Dewey, Solomon. *A Short and Easy Method of Surveying.* Hartford: Elisha Babcock, 1799. Evans: 35406 Karpinsky: p. 122. Surveying textbook.

705 Hauducoeur, C.P. *A Map of the Head of Chesapeake Bay and Susquehanna River, Shewing the Navigation of the Same with a Topographical Description of the Surrounding Country from an Actual Survey.* [Philadelphia: Thomas Dobson and William Cobbett], 1799. Not included in Readex Wheat/Brun: 520. Map of the area at the mouth of the Susquehanna and inland; gives property lines and names of owners as well as roads, settlements, and soundings.

706 Jess, Zachariah. *The American Tutor's Assistant, Improved.* Wilmington: Bonsal and Niles, 1799. Evans: 35669 Karpinsky: p. 123. Arithmetic textbook; Jess is described as a "schoolmaster in Wilmington." There were several later editions, the last in 1830.

707 Jess, Zachariah. *A Compendious System of Practical Surveying and Dividing of Land.* Wilmington: Bonsal and Niles, 1799. Evans: 35670 Karpinsky: p. 124. Surveying textbook.

708 Latrobe, Benjamin Henry (1764-1820). *Remarks on the Address of the Committee of the Delaware and Schuylkill Canal Company to the Committee of the Senate and House of Representatives As Far As It Notices the "View of Practicality and Means of Supplying the City of Philadelphia with Wholesome Waters."* Philadelphia:

Zachariah Poulson, Jr., 1799. Evans: 35714. Latrobe's defense of his plan to bring Schuylkill water to Philadelphia (see 709).

709 Latrobe, Benjamin Henry (1764-1820). *View of the Practicality and Means of Supplying the City of Philadelphia with Wholesome Water.* Philadelphia: Zachariah Poulson, Jr., 1799. Evans: 35714. Plan to bring water from the Schuylkill River to Philadelphia; includes engineering plans. Latrobe's idea for a city water supply was the first in America and was inspired by the Philadelphia yellow fever epidemics. According to the DAB "[Latrobe's] plan was so practical that it led to the immediate abandonment of the other schemes." Latrobe was engineer for the subsequent construction, and the waterworks he built operated successfully until 1815 when it was no longer adequate and was superseded by a new design.

710 Little, Ezekiel (1762-1840). *The Usher.* Exeter: H. Ranlet, 1799. Evans: 35734 Karpinsky: p. 125. Combined arithmetic and surveying textbook.

711 *Map of the State of Delaware and Eastern Shore of Maryland.* Philadelphia: [1799]. Evans: 35768 Not included in Readex Wheat/Brun: 486. Only copy is missing.

712 Massachusetts Society for Promoting Agriculture. *Papers on Agriculture, Consisting of Communications Made to the Massachusetts Society for Promoting Agriculture with Extracts from Various Publications.* Boston: Young and Minns, 1799. Evans: 35802. Includes a list of premiums, articles on crops, trees, pest control, and animal husbandry. Authors include (no first names given) Peck, Winthrop, Russell, Cooper, Wadsworth, Edwards, Kirwan, Hillhouse, Welles, Ure, Putnam, and Clarke.

713 New York Society for the Promotion of Agriculture, Arts, and Manufactures. *Transactions of the New York Society for the Promotion of Agriculture, Arts, and Manufactures.* Albany: Loring Andrews, 1799. Evans: 35935. Includes essays on crop cultivation, plant diseases, fertilizer, livestock, manufacturing, construction, and mineralogy. Authors include Livingston, L'Hommedieu, Mitchill, Woodhouse, John Watkins, Shadrach Ricketson, Robert Johnson, Benjamin Folger, John Stevens, Simeon DeWitt, and Andrew Billings.

714 Payne, John (fl. 1800). *A New and Complete System of Universal Geography. Vol. IV.* New York: John Low, 1799. Evans: 36047. Geography textbook.

715 [Peck, John (1735-1812)]. *Facts and Calculations Respecting the Population and Territory of the United States of America.* [Boston:1799]. Evans: 35051. Calculations about the population of United States based on immigration statistics, among other factors; also calculations about the area of the country in acres, occupied and unoccupied.

716 Philadelphia Joint Committee on Water Supply. *Report to the Select and Common Councils on the Progress and State of the Water Works.* Philadelphia: Zachariah Poulson, Jr., 1799. Evans: 36091. History of the city water project, particularly implementation of Latrobe's plan (see 709); includes a description of construction completed at the time.

717 Prentiss, Thomas Mellen (1773-1823). *The Maine Spelling Book.* Leominster, MA: Charles and John Prentiss, 1799. Evans: 36156 Karpinsky: p. 126. Textbook including a "geographical description" of Maine.

718 Scott, Joseph. *The New and Universal Gazetteer or Modern Geographical Dictionary. Vols. 1 and 2.* Philadelphia: Francis and Robert Bailey, 1799. Evans: 36282. Geographical encyclopedia including astronomy.

719 *A Statistical Table for the United States of America for a Succession of Years, Compiled Chiefly from Official Documents.* [Philadelphia:1799.] Evans: 36262. Table of statistics including population, improved land, militia and navy, domestic produce, revenue, ship tonnage, expenditures, money, and debt.

720 [Varle, Charles]. *A Plan of the City and Environs of Baltimore, Respectfully Dedicated to the Mayor, City Council, and Citizens Thereof by the Author.* Philadelphia: 1799. Not included in Readex Wheat/Brun: 526. Topographical map of Baltimore and environs; includes names of owners of "outlying residences" (Wheat and Brun).

721 [Williams, Jonathan (1750-1815)]. *Thermometrical Navigation.* Philadelphia: R. Aitken, 1799. Evans: 36722. Subtitled "A series of experiments and observations tending to prove that by ascertaining the relative heat of the seawater from time to time, the passage of a ship through the gulph [sic] stream and from deep water into soundings may be discovered in time to avoid danger. . . ." Includes extracts from Franklin's *Maritime Observations* (see 480) and Williams' memoir on the use of the thermometer in navigation (see 548), along with tables of measurements and observations from others in an appendix.

722 [Williamson, Charles]. *Description of the Settlement of the Genesee Country.* [1799]. Evans: 36727. Promotional pamphlet; Williamson was a land speculator and promoter.

Natural and Physical Science Titles

1665

723 Danforth, Samuel (1626-1674). *An Astronomical Description of the Late Comet or Blazing Star As It Appeared in New England in the 9th, 10th, 11th and the Beginning of the 12th Moneth [sic], 1664, Together with a Brief Theological Application Thereofe [sic].* Cambridge: Samuel Green, 1665. Evans: 99. Description of a comet sighting; Danforth uses sermon format (he was a minister at Roxbury, Massachusetts).

1683

724 Mather, Increase (1639-1723). *Kometographia or a Discourse Concerning Comets; Wherein the Nature of Blazing Stars Is Enquired into, with an Historical Account of All the Comets Which Have Appeared from the Beginning of the World unto This Present Year M.DC.LXXIII, Expressing the Place in the Heavens Where They Were Seen, Their Motions, Forms, Duration; and the Remarkable Events Which Have Followed in the World, So Far As They Have Been by Learned Men Observed, As Also Two Sermons Occasioned by the Late Blazing Stars.* Boston: S. Green, 1683. Evans: 352. Summary of theories about comets as well as a history of comet sightings. Mather had viewed Halley's comet in 1682 at Harvard; Stearns describes this as an "erudite study," showing Mather's knowledge of European sources. Mather views comets as natural phenomena, but he also asserts their function as indicators of calamity.

1719

725 Prince, Thomas (1687-1758). *An Account of a Strange Appearance in the Heavens on Tuesday-Night, March 6, 1716, As It Was Seen over Stow-Market in Suffolk in England.* Boston: S. Kneeland, 1719. Evans: 2068. Description of Prince's observation of "great streams of smoakey light" in the sky; in the form of a letter to Francis Prince in London.

726 Robie, Thomas (1689-1729). *A Letter to a Certain Gentleman Desiring a Particular Account May Be Given of a Wonderful Meteor That Appeared in New England, on Decemb. 11.1719. in the Evening.* Boston: J. Franklin, 1719. Evans: 2171. Description of Robie's observation of aurora borealis; written to William Derham of London. Stearns calls the account "remarkable" because Robie regards the aurora as a natural phenomenon and tries to explain it "by reference to John Wallis' nitrous-sulphureous theories."

1721

727 Mather, Cotton (1662-1728). *The World Alarm'd, a Surprizing [sic] Relation of a New Burning-Island Lately Raised out of the Sea, Near Tercera; with a Geographical and Theological Improvement of So Astonishing an Occurrence and a Brief History of* *the Other Ignivomous [sic] Mountains at This Day Flaming in the World, in a Letter to an Honourable Fellow of the Royal Society at London from a Member of the Same Society at Boston.* Boston: B. Green, 1721. Evans: 2255. Account of a new volcanic island in the Azores near Terceira. Mather sent the account to the Royal Society along with drawings provided by a sea captain and a piece of pumice from the volcano.

1726

728 Greenwood, Isaac (1702-1745). *An Experimental Course of Mechanical Philosophy Whereby Such a Competent Skill in Natural Knowledge May Be Attained to (by Means of Various Instruments and Machines, with Which There Are Above Three Hundred Curious and Useful Experiments Performed) That Such Persons As Are Desirous Thereof May, in a Few Weeks Time, Make Themselves Better Acquainted with the Principles of Nature and the Wonderful Discoveries of the Incomparable Sir Isaac Newton Than by a Year's Application to Books and Schemes.* Boston: 1726. Evans: 2746 Karpinsky: p. 43. Description in outline form of a series of sixteen public lectures and demonstrations, the first in America on Newton's theories. These lectures were probably instrumental in bringing about Greenwood's appointment as the first Hollis Professor of Mathematics and Natural Philosophy at Harvard in 1727.

729 Greenwood, Isaac (1702-1745). *A Course of Philosophical Lectures with a Great Variety of Curious Experiments Illustrating and Confirming Sir Isaac Newton's Laws of Matter and Motion.* [Boston?: c.1726]. mp: 39848 Bristol: 715 Karpinsky: p. 41. Another lecture series outline; includes lectures on mechanics and statics.

1727

730 Mather, Cotton (1662-1728). *Boanerges: A Short Essay to Preserve and Strengthen the Good Impressions Produced by Earthquakes on the Minds of People That Have Been Awakened with Them, with Some Views of What Is to Be Further and Quickly Look'd For.* Boston: S. Kneeland, 1727. Evans: 2908. Sermon with appendix describing all earthquakes occurring internationally during 1727.

731 Mather, Cotton (1662-1728). *The Terror of the Lord; Some Account of the Earthquake That Shook New England in the Night between the 29 and the 30 of October, 1727, with a Speech Made unto the Inhabitants of Boston, Who Assembled the Next Morning for the Proper Exercises of Religion, on So Uncommon and So Tremendous an Occasion.* Boston: T. Fleet, 1727. Evans: 2921 Hazen: 6867. Account of the earthquake of 1727, including its sound, duration, and violence. An appendix describes the geographical extent of the earthquake, its aftershocks, and past New England earthquakes. It includes Mather's post-earthquake sermon.

732 Prince, Thomas (1687-1758). *Two Sermons on Psal. XVIII 7, at the Particular Fast in Boston, Nov. 2, and the General Thanksgiving, Nov. 9, Occasioned by the Late Dreadful Earthquake.* Boston: D. Henchman, 1727. Evans: 2945. Sermon including a brief account of current theories concerning the cause of earthquakes. It was reissued in 1755 with an appendix suggesting that the cause of earthquakes was electrical and that Bostonians might have caused a 1755 earthquake by installing lightning rods. This theory was disputed by John Winthrop (see 744).

1731
733 Greenwood, Isaac (1702-1745). *A Philosophical Discourse Concerning the Mutability and Changes of the Material World.* Boston, 1731. Evans: 3426 Not included in Readex Karpinsky: p. 45. Lecture delivered upon the death of Thomas Hollis; Stearns describes it as expounding "Newtonianism" and suggests it indicates Greenwood's rejection of an "apocalyptic view of the world" in favor of a "mechanistic explanation."

1734
734 Greenwood, Isaac (1702-1745). *Explanatory Lectures on the Orrery, Armillary Sphere, Globes and Other Machines, Instruments, and Schemes Made Use of by Astronomers: Accompanied with a Great Variety of Physical Experiments and Curious Remarks.* Boston: 1734. Evans: 3776 Karpinsky: p. 48. Outline of a lecture series inspired by a gift of scientific apparatus to Harvard by Thomas Hollis' nephew.

1737
735 Cooper, William (1694-1743). *Concio Hyemalis, a Winter Sermon, Being a Religious Improvement of the Irresistible Power of God's Cold.* Boston: J. Draper, 1737. Evans: 4134 Primarily sermon on winter, but it contains explanations of cold weather phenomena taken from Robert Boyle and Cotton Mather.

1743
736 Franklin, Benjamin (1706-1790). *A Proposal for Promoting Useful Knowledge among the British Plantations in America.* Philadelphia: Benjamin Franklin, 1743. Evans: 5190 Miller: 325. Proposal for founding the American Philosophical Society; although it was issued by Franklin's press over Franklin's signature, Hindle suggests it was a "joint project" of several interested parties, including John Bartram. Hindle also suggests that the subject matter proposed for the society is slanted more towards natural than physical science, reflecting the interests of the "natural history circle" of which Franklin was a part.

1744

737 *An Essay on Comets, Their Nature, the Laws of Their Motions, the Cause and Magnitude of Their Atmosphere and Tails; with a Conjecture of Their Use and Design.* Boston: Rogers and Fowle, 1744. Evans: 5389. Discussion of comets based on Newtonian theories; includes history and a table of astronomical measurements.

738 *Just Arrived from London, for the Entertainment of the Curious and Others, and Is Now to Be Seen by Six or More, in a Large Commodious Room at the House of Mr. Vidal in Second-Street: The Solar or Camera Obscura Microscope, Invented by the Ingenious Dr. Liberkhun.* [Philadelphia: Benjamin Franklin, 1744.] Evans: 5419 Miller: 353. Broadside advertisement for a combination microscope and "camera obscura"; includes a description of the phenomena to be shown.

1745

739 **Colden, Cadwallader (1688-1776).** *An Explication of the First Causes of Action in Matter and of the Cause of Gravitation.* New York: James Parker, 1745. Evans: 5564. Colden's attempt to explain the "cause of gravitation," based chiefly on his reading of Newton's "General Scholium" in the *Principia* and the "Queries" at the end of *Optics.* It appeared in several European editions, but met with a confused and largely negative response. The BDAS calls it the "most ambitious scientific undertaking in [the] American colonial period," but Stearns describes it as "rational, non-experimental and ill-informed."

1749

740 **Prince, Thomas (1687-1758).** *The Natural and Moral Government and Agency of God in Causing Droughts and Rains.* Boston: Kneeland and Green, 1749. Evans: 6408. Sermon containing scientific descriptions of various natural phenomena; includes references to Newton, Halley, Boyle, and Desagulier, among others.

1752

741 **Kinnersley, Ebenezer (1711-1778).** *Newport, March 16, 1752: Notice Is Hereby Given to the Curious That at the Courthouse in the Council Chamber Is Now to Be Exhibited and Continued from Day to Day, for a Week or Two, a Course of Experiments on the Newly Discovered Electrical Fire.* [Newport:1752]. mp: 40620 Bristol:1572. Notice of a lecture/ demonstration by Franklin's associate, Ebenezer Kinnersley. Lemay describes this lecture as "extraordinarily successful in meeting the religious prejudice of his audience."

1753

742 **[Alexander, James (1691-1756)].** *Letters Relating to the Transit of Mercury over the Sun, Which Is to Happen May 6, 1753.* Philadelphia: 1753. Evans: 7038. Translation of letters written by Joseph-Nicolas Delisle of the Academie des Sciences concerning the

observation of the Transit of Mercury. This translation was distributed
throughout the colonies by Franklin and Alexander in hope of
"stimulating observations" (Stearns).

1755

743 Winthrop John (1714-1779). *A Lecture on Earthquakes.*
Boston: Edes and Gill, 1755. Evans: 7597 Hazen: 11002. Winthrop's
response to the earthquake of 1755; it is an attempt to quiet public fears
by describing earthquakes as natural phenomena rather than divine
intervention. According to Hindle, Winthrop's "recognition of the wave
character of earthquakes" preceded the later statement by John Mitchell,
who is usually given credit for the concept.

1756

744 Winthrop, John (1714-1779). *A Letter to the Publishers of the
Boston Gazette, &c. , Containing an Answer to the Rev. Mr. Prince's
Letter, Inserted in Said Gazette on the 26th of January, 1756.* Boston:
1756. Evans: 7820 Hazen: 11003. Winthrop's answer to Thomas
Prince's assertion that the cause of earthquakes was electrical (see 732).
Prince's reaction to Winthrop's *Lecture on Earthquakes* (743) sparked a
lively debate between them; Jared Eliot asserted in the end that Winthrop
had "laid Mr. Prince flat on [his] back" (Hindle).

1759

745 Winthrop, John (1714-1779). *Two Lectures on Comets, Read
in the Chapel of Harvard College in Cambridge, New England in April,
1759, on Occasion of the Comet Which Appear'd in That Month.*
Boston: Green and Russell, 1759. Evans: 8522. Discussion of theories
regarding comets, with some speculation; inspired by the return of
Halley's comet in 1759. Hindle calls it "a precise piece of careful
cogitation" which led Winthrop to further calculations concerning the
mass and density of comets, later submitted to the Royal Society.

1761

746 [Perkins, John (1698-1781)]. *An Essay on the Agitations of
the Sea and Some Other Remarkables Attending the Earthquakes of the
Year M,DCC,L,V, to Which Are Added, Some Thoughts on the Causes
of Earthquakes, Written in the Year 1756.* Boston: B. Mecom, 1761.
Evans: 8851 Hazen: 3605. Discussion of the relationship between
earthquakes and disturbances at sea; includes a discussion of
topography. Perkins suggests that settling land is the cause of
earthquakes.

747 Winthrop, John (1714-1779). *Relation of a Voyage from
Boston to Newfoundland for the Observation of the Transit of Venus,
June 6, 1761.* Boston: Edes and Gill, 1761. Evans: 9040. Winthrop's
first account of his observations of the Transit of Venus at St. John's,
Newfoundland. Winthrop explains the usefulness of the observations,

the procedures involved and his calculations. A copy was sent to the Royal Society.

1764

748 Kinnersley, Ebenezer (1711-1778). *A Course of Experiments in That Curious and Entertaining Branch of Natural Philosophy Called Electricity; Accompanied with Explanatory Lectures in Which Electricity and Lightning Will Be Proved to Be the Same Thing.* [Philadelphia]: Armbruster, 1764. Evans: 9708. Syllabus of a course of lectures which Kinnersley delivered at the College of Philadelphia (University of Pennsylvania), where he was a professor of oratory and rhetoric. According to the BDAS, Kinnersley's ability as a speaker and demonstrator helped spread the news about the Philadelphia experiments in electricity. Stearns calls him "the most able scientist to give lectures in the colonies" after Isaac Greenwood.

1765

749 Johnson, William (d. 1768). *A Course of Experiments in That Curious and Entertaining Branch of Natural Philosophy Call'd Electricity; Accompanied with Lectures on the Nature and Properties of the Electrical Fire.* New York: H. Gaine, 1765. Evans: 10027. Syllabus of lectures by a competitor of Kinnersley; content is similar to 748. According to Lemay, Johnson was "the second most famous lecturer on electricity in America" after Kinnersley.

750 Mason, David. *A Course of Experiments in That Instructive and Entertaining Branch of Natural Philosophy Called Electricity, to Be Accompanied with Methodical Lectures on the Nature and Properties of This Wonderful Element.* [Boston: 1765]. mp: 41559 Bristol: 2594. Another series of lectures on electricity.

1769

751 West, Benjamin (1730-1813). *An Account of the Observations of Venus upon the Sun, the Third Day of June, 1769, at Providence in New England, with Some Account of the Use of Those Observations.* Providence: John Carter, 1769. Evans: 11525. Description of the Providence observations of the Transit of Venus, along with explanations of the transit's importance and a history of past observations. West was an almanac-maker and later professor of mathematics and astronomy at Brown University; these Providence observations were reprinted in the *Transactions* of the American Philosophical Society.

752 Winthrop, John (1714-1779). *Two Lectures on the Parallax and Distance of the Sun, As Deducible from the Transit of Venus.* Boston: Edes and Gill, 1769. Evans: 11536. Winthrop's explanation of the method and purposes of his observations of the Transit of Venus in 1769; intended for the public and his students at Harvard. Hindle calls it "the finest piece of descriptive writing on the transit produced in

America," using "prose that was characteristically clear, concise, and coherent."

1770

753 Rush, Benjamin (1745-1813). *A Syllabus of a Course of Lectures on Chemistry for the Use of the Students of Medicine in the College of Philadelphia.* Philadelphia: Charles Cist, 1770. Evans: 18173 Austin: 1686 Guerra: a-715 Hazen: 91318. Chemistry text presented in outline form; first chemistry text written by an American. According to D'Elia, Rush followed a system developed by Dr. Joseph Black, under whom Rush studied at Edinburgh, and according to Hawke, had the lectures themselves been published (rather than this outline), they "would have been condemned as plagiarism" of Black's work.

1771

754 American Philosophical Society. *Transactions of the American Philosophical Society Held at Philadelphia for Promoting Useful Knowledge. Volume I, from January 1st, 1769 to January 1st, 1771.* Philadelphia: William and Thomas Bradford, 1771. Evans: 11959 Meisel: 2: p. 6. Largely taken up with papers on the Transit of Venus and other astronomical material; Sections II-IV include articles on agriculture, botany, geography, engineering, chemistry, and medicine. Contributors include David Rittenhouse, William Smith, Benjamin West, Nevil Maskelyne, Hugh Williamson, Thomas Bond, John Morgan, Benjamin Rush, Samuel Bard, Thomas Gilpin, and John Jones among others. Another edition appeared in 1789 (Evans: 21651).

1772

755 Oliver, Andrew Jr. (1731-1799). *An Essay on Comets in Two Parts.* Salem: Samuel Hall, 1772. Evans: 12498. Analysis of comet tails, including Oliver's hypothesis that comets "may be inhabited worlds." The work was highly regarded by French astronomer Jean-Sylvain Baily, who published a translation in France.

756 Smith, William (1727-1803). *An Oration, Delivered January 22, 1773, before the Patron, Vice-Presidents and Members of the American Philosophical Society Held at Philadelphia for Promoting Useful Knowledge.* Philadelphia: John Dunlap, 1772. Evans: 13022. Smith's description of the aims of the American Philosophical Society, emphasizing the need for both "those Branches of Literature and Science whereby the mind may be humanized," as well as "useful sciences" which will save labor and "multiply the conveniences of life." Smith was a founding member of the Society and Provost of the College of Philadelphia (University of Pennsylvania).

1773

757 Colles, Christopher (1738-1816). *Syllabus of a Course of Lectures in Natural Experimental Philosophy.* Philadelphia: John Dunlap, [1773]. Evans: 12730 Guerra: a-515. Outline of eighteen lectures on mechanics, "pneumatics," hydraulics, "hydrostatics," geography, and the globe. Colles was an engineer and an inventor.

758 Rush, Benjamin (1745-1813). *Experiments and Observations on the Mineral Waters of Philadelphia, Abington, and Bristol, in the Province of Pennsylvania.* Philadelphia: James Humphreys, 1773. Evans: 12995 Austin: 1642 Guerra: a-534. Description of a series of analytical tests of various springs and suggestions for medical use of the waters. It was inspired by John De Normandie's analysis of spring water from Bristol in the *Transactions* of the American Philosophical Society.

1775

759 Rittenhouse, David (1732-1796). *An Oration, Delivered February 24, 1775, before the American Philosophical Society Held at Philadelphia for Promoting Useful Knowledge.* Philadelphia: John Dunlap, 1775. Evans: 14432. Discussion of astronomy, including history, the most important discoveries, and current problems. Rittenhouse was one of the founders of the American Philosophical Society and one of its first presidents.

1780

760 American Academy of Arts and Sciences. *An Act to Incorporate and Establish a Society for the Cultivation and Promotion of Arts and Sciences.* Boston: Benjamin Edes, 1780. Evans: 16841 Hazen: 6849. By-laws and purpose of the Academy.

761 Bowdoin, James (1726-1790). *A Philosophical Discourse, Addressed to the American Academy of Arts and Sciences, in the Presence of a Respectful Audience Assembled at the Meeting-House in Brattle Street in Boston, on the Eighth of November, 1780, after the Inauguration of the President into Office.* Boston: Benjamin Edes, 1780. Evans: 16720. Address delivered after Bowdoin's inauguration as the first president of the American Academy of Arts and Sciences. Bowdoin surveys the Academy's fields of interest and ends with an encomium to Harvard.

1781

762 Clap, Thomas (1703-1767). *Conjectures upon the Nature and Motion of Meteors, Which Are above the Atmosphere.* Norwich: John Trumbull, 1781. Evans: 17113. Clap's hypotheses about the orbits of meteors and their similarities to comets. The work was published posthumously by Ezra Stiles; it was disputed by John Winthrop.

1782

763   Bond, Thomas (1712-1784). *Anniversary Oration, Delivered May 21st, before the American Philosophical Society Held in Philadelphia for the Promotion of Useful Knowledge.* Philadelphia: John Dunlap, 1782. Evans: 17479. Short history of the American Philosophical Society, followed by a discussion of "The Rank and Dignity of Man in the Scale of Being," an attempt to rank human beings in relation to other animals.

764   [Macpherson, John]. *An Introduction to the Study of Natural Philosophy.* [Philadelphia: 1782]. Evans: 17582 Hazen: 6652. Outline of a series of lectures on natural philosophy.

1783

**765 ** *A New System of Philosophy or the Newtonian Hypothesis Examined by an American.* Poughkeepsie: John Holt, 1783. Evans: 18058. Examination of Newton's theories in light of Franklin's discoveries; strongly nationalistic, asserting that Franklin has contradicted and superseded Newton. The Readex copy has "A Discourse on Burns and Scalds" inserted in the middle of the discussion.

1784

**766 ** *An Accurate and Complete Description of Sleep, in a Discourse Delivered before the Medical Society, September, 1782.* New York: 1784. Evans: 18318 Austin: 17. Argument for the existence of souls, citing Newton, among others. The author claims it is a response to Ledyard (see 767).

767   [Ledyard, Isaac (1754-1803)]. *An Essay on Matter.* Philadelphia: 1784. Evans: 18554. Materialistic description of physical processes; Hindle describes it as "a rambling account . . . which demonstrated what could happen when the uninformed accepted the idea that the 'book of nature' was open for all to read."

768   Moyes, Henry. *Heads of a Course of Lectures on the Natural History of the Celestial Bodies, the Earth, the Vegetable, the Atmosphere, and Animal Kingdoms.* [Boston?: c.1784]. mp: 44559 Bristol: 5927. Outline of a public lecture series. Moyes was a professional lecturer from England who carried out a popular series of lectures in American cities during 1784-1785.

769   Moyes, Henry. *Heads of a Course of Lectures on the Philosophy of Chemistry and Natural History.* [Boston?: c.1784]. mp: 44560 Bristol: 5928. Outline of public lecture series (see 768).

770   Strong, Nehemiah (1729-1807). *Astronomy Improved, or a New Theory of the Harmonious Regularity Observable in the Mechanism or Movements of the Planetary System.* New Haven:

Thomas and Samuel Green, 1784. Evans: 18797. Discussion of
astronomical principals; Strong was professor of mathematics and
natural philosophy at Yale. According to Hindle, "this elementary
account failed to present anything new and even indicated that Strong
did not fully understand the old system."

1785

771 American Academy of Arts and Sciences. *Memoirs of the
American Academy of Arts and Sciences to the End of the Year
M,DCC,LXXXIII. Vol.1.* Boston: Adams and Nourse, 1785. Evans:
18900 Meisel: 2: p. 40. First collection of papers submitted to the
Academy. Topics include astronomy, physics, geology, meteorology,
agriculture, and medicine. Contributors include James Bowdoin,
Arthur Lee, Joseph Willard, Stephen Sewall, Benjamin West, Manasseh
Cutler, James Winthrop, Jeremy Belknap, and Samuel Deane among
others.

772 Marshall, Humphrey (1722-1801). *Arbustrum Americanum:
The American Grove, or an Alphabetical Catalogue of Forest Trees and
Shrubs, Natives of the United States, Arranged According to the
Linnean System.* Philadelphia: Joseph Crukshank, 1785. Evans: 19068
Meisel: 3: p. 354. Alphabetical listing of American trees and shrubs
using Linnean classifications. Hindle calls it "the first systematic
botanical book to appear in the United States." Marshall used
information on American botany from several sources, including his
own observations; he also included a discreet advertisement for his own
seeds and plants. The work was translated in Leipzig and Paris.

773 Vancouver, Charles (fl. 1785-1813). *General Compendium
or Abstract of Chemical, Experimental, and Natural Philosophy in Four
Volumes. Vol. 1.* Philadelphia: 1785. Evans: 19337. Begins with an
ambitious table of contents for a four-volume set, but only one volume
was published. Volume I covers the Universe, the Sun, and Heat and
Cold; it includes one plate of a "Machine for Preserving from Fire,"
showing people being lowered in a bucket from a burning building.

1786

774 American Philosophical Society. *Transactions of the American
Philosophical Society Held at Philadelphia for Promoting Useful
Knowledge. Vol. 2.* Philadelphia: Robert Aitken, 1786. Evans:
19465. Subjects include natural history, meteorology, astronomy,
physics, agriculture, medicine, and engineering. Contributors include
Benjamin Franklin, David Rittenhouse, Jeremy Belknap, Thomas
Hutchins, Andrew Oliver, John Morgan, and Bernard Romans among
others.

775 Bowdoin, James (1727-1790). *A Philosophical Discourse
Addressed to the American Academy of Arts and Sciences, to Which
Are Added Three Memoirs on Philosophical Subjects.* Boston: Adams

and Nourse, 1786. Evans: 19520. Extracted from the *Memoirs* (see 771); suggests subjects for future study. The three "Memoirs" are "Observations upon an Hypothesis for Solving the Phenomena of Light," "Observations on Light and the Waste of Matter in the Sun and Fixt [sic] Stars," and "Observations Tending to Prove . . . the Existence of an Orb Which Surrounds the Whole Visible Material System."

776 Ladd, Joseph Brown (1764-1786). *An Essay on Primitive, Latent, and Regenerated Light.* Charleston: Bowen and Markland, [1786]. Evans: 19746. Description of the composition of light, with attempts to "account for every luminous phenomenon. . . ." Ladd is best known as a minor poet; Hindle calls this essay "a strange and unprofitable performance."

777 *Meteorological Observations Made at Springmill, Thirteen Miles NNW of Philadelphia, 40° 9' N.* [Philadelphia?: 1786]. mp: 44924 Bristol: 6322. Table of data giving temperature, barometric readings, wind direction, precipitation, and weather conditions for various dates.

1787

778 Carleton, Osgood (1742-1816). *By Permission, Mr. Carleton, Professor of Astronomy, Proposes (with the Approbation of the Ladies and Gentlemen of This Metropolis) to Deliver a Course of Five Lectures on That Sublime Science.* Boston: E. Russell, [1787]. Evans: 20261. Outline of a series of public lectures on astronomy.

779 Rush, Benjamin (1745-1813). *Syllabus of Lectures, Containing the Application of the Principles of Natural Philosophy and Chemistry to Domestic and Culinary Purposes.* Philadelphia: Andrew Brown, 1787. mp: 45160 Bristol: 6586. Outline of lectures "composed for the use of the Young Ladies Academy." The DSB describes the series as a reflection of Rush's belief that "chemistry should be useful to the larger community."

780 Smith, Samuel Stanhope (1750-1819). *An Essay on the Causes of the Variety of Complexion and Figure in the Human Species.* Philadelphia: Robert Aitken, 1787. Evans: 20712. Speech originally delivered to the American Philosophical Society. Smith argues for a single origin for all races, suggesting that racial differences were caused by climate and "the state of society." Greene cites this essay as the first anthropological treatise written in America; it was republished in London and Edinburgh and reissued in an enlarged edition in 1810.

1788

781 *Meteorological Observations Made at Springmill, Thirteen Miles NNW of Philadelphia, 40° 9' N.* [Philadelphia?: 1788]. mp: 45296

Bristol: 6739. Table of data giving temperatures, barometric readings, wind directions, precipitation, and weather conditions for several dates.

782 Waterhouse, Benjamin (1754-1846). *Heads of a Course of Lectures Intended As an Introduction to Natural History.* Providence: Bennett Wheeler, [c.1788]. mp: 45406 Bristol: 6859. Outline of lecture series offered first at the College of Rhode Island (Brown University); Hindle describes them as "widely influential in developing a taste for natural history."

1789

783 Burges, Bartholomew. *A Short Account of the Solar System, and of Comets in General, Together with a Particular Account of the Comet That Will Appear in 1789.* Boston: B. Edes and Son, 1789. Evans: 21721. Description of the solar system and comets, with a discussion of Halley's calculations for the comet's orbit. It was written against the background of the return of Halley's comet in 1788-89.

1790

784 College of Rhode Island [Brown University]. *The Reverend Peres Forbes, Professor of Natural and Experimental Philosophy, in Rhode Island College, Proposes to Exhibit a Course of Lectures upon Natural Philosophy and Astronomy.* Providence: 1790. Evans: 22851. Description of a lecture series covering botany, agriculture, mechanics, hydraulics, pneumatics, optics, astronomy, electricity, magnetism, and anatomy.

785 *Encyclopaedia, or a Dictionary of Arts, Sciences, and Miscellaneous Literature.* Philadelphia: Thomas Dobson, 1790. Evans: 22486. First American edition of the *Encyclopedia Britannica.* According to Evans, although based on the British edition, the American edition included new sections on "America" (by Jedidiah Morse), "Anatomy," and "Chemistry." In addition 543 new copperplates were executed by American engravers.

786 Penington, John (1768-1793). *Chemical and Economical Essays, Designed to Illustrate the Connection between the Theory and Practice of Chemistry, and the Application of That Science to Some of the Arts and Manufactures of the United States of America.* Philadelphia: Joseph James, 1790. Evans: 22757 Hazen: 8204. Chemistry text describing chemical processes and theories with an emphasis on the use and derivation of dyes and other industrial processes. Penington was a former student of Benjamin Rush who organized the first chemical society in the United States (1789).

787 Penington, John (1768-1793). *An Inaugural Dissertation on the Phoenomena, Causes and Effects of Fermentation.* Philadelphia: Joseph James, 1790. Austin: 1479. Evans: 22758. Medical dissertation for the University of Pennsylvania. Penington describes the process of fermentation and distinguishes between it and other, similar processes.

1791

788 Bartram, William (1739-1823). *Travels through North and South Carolina, Georgia, East and West Florida, the Cherokee Country, the Extensive Territories of the Muscogulges or Creek Confederacy, and the Country of the Choctaws.* Philadelphia: James and Johnson, 1791. Evans: 23159 Meisel: 3: p. 356. Description of the natural history and geography of the region, including a "catalogue of birds of North America." Bartram's major work, it was republished widely in the United States and Europe and influenced Benjamin Smith Barton, among others. Bartram was the son of John Bartram (see 39).

789 Reiche, Charles Christopher. *Fifteen Discourses on the Marvelous Works in Nature, Delivered by a Father to His Children.* Philadelphia: James and Johnson, 1791. Evans: 23729 Meisel: 3: p. 357. Textbook of natural history covering such topics as meteorology, botany, zoology, and anatomy with a heavy religious orientation. The book begins with letters of commendation from Rittenhouse, Barton, and Rush, among others.

790 *A System of Chemistry: Comprehending the History, Theory, and Practice, According to the Latest Discoveries and Improvements.* Philadelphia: Thomas Dobson, 1791. Evans: 23817. Extracted from Dobson's edition of the *Encyclopedia Britannica* (see 785); covers history, theory, and practice.

1792

791 American Philosophical Society. *At a Meeting of the Committee.* [Philadelphia: 1792]. mp: 46375 Bristol: 7920. Report of a committee charged with developing the natural history of and control methods for the "hessian-fly." The report is a collection of research along with nine questions to be pursued. Committee members were Jefferson, Barton, James Hutchinson, and Caspar Wistar.

792 Belknap, Jeremy (1744-1798). *A Discourse Intended to Commemorate the Discovery of America by Christopher Columbus.* Boston: Belknap and Hall, 1792. Evans: 24085. Includes a discussion of "Whether the honey-bee is a native of America." Belknap argues that it is, against arguments by Jefferson and others that bees were imported from Europe.

793 Mitchill, Samuel Latham (1764-1831). *Outlines of the Doctrines in Natural History, Chemistry, and Economics.* New York: Childs and Swaine, 1792. Evans: 24549 Hazen: 7268 Meisel: 3: p. 358. Syllabus of a course of lectures given at the College of New York (Columbia). Mitchill was appointed professor of natural history, chemistry, and agriculture in 1792.

794 Pope, John (1770-1845). *A Tour Through the Southern and Western Territories of the United States of North-America; the Spanish Dominions on the River Mississippi and the Floridas; the Countries of the Creek Nations and Many Uninhabited Parts.* Richmond: John Dixon, 1792. Evans: 24705 Meisel: 3: p. 358. Travel narrative including some natural history of the region.

795 [Riley, George]. *The Beauties of the Creation or a New Moral System of Natural History Displayed in the Most Singular, Curious, and Beautiful Quadrupeds, Birds, Insects, Trees, and Flowers.* Philadelphia: William Young, 1792. Evans: 24745 Meisel: 3: p. 358. Natural history textbook with heavy religious orientation. Riley describes his purpose as being to "elevate the heart to the creator" rather than to present detail studied by the "mere speculatist."

796 Rouelle, John [Jean] (b. 1751). *A Complete Treatise on the Mineral Waters of Virginia, Containing a Description of Their Situation, Their Natural History, Their Analysis, Contents and Their Use in Medicine.* Philadelphia: Charles Cist, 1792. Evans: 24757 Austin: 1625 Hazen: 9040. Analytical study; Rouelle's interest is chiefly medical, but he does include some account of his experiments with reagents. Rouelle was a French chemist appointed mineralogist-in-chief and professor of natural history, chemistry, and botany at the Academie des Sciences et Beaux-Arts des Etats-Unis (Richmond, VA). He formed a collection of animals, plants, and minerals which he transported to France in 1788.

797 *A Small Sketch on Natural Philosophy.* Wilmington: Brynberg and Andrews, 1792. Evans: 25576. Description of the laws of physics, astronomy, and geography. The author suggests that the planets are inhabited. The orientation is heavily religious.

798 Wilkins, Henry (1767-1847). *An Original Essay on Animal Motion, in Which the Instrument Thereof, Its Definition, and Mode of Operating Are Minutely Treated upon.* Philadelphia: 1792. Evans: 25036 Austin: 2054. Somewhat muddled treatise on the origin of motion and its effects; Wilkins suggests that "animal motion" is the "basis of physic."

1793

799 American Academy of Arts and Sciences. *Memoirs. Vol II.* Boston: Isaiah Thomas and Ebenezer T. Andrews, 1793. Evans: 25092. Contains papers on mathematics, astronomy, physics, meteorology, biology, agriculture, manufacturing, chemistry, and medicine. Authors include James Winthrop, Joseph Willard, Edward Holyoke, Caleb Gannett, John Vinall, and Noah Webster, among others.

800 American Philosophical Society. *Transactions of the American Philosphical Society, Held at Philadelphia for Promoting Useful*

Knowledge, Vol. III. Philadelphia: Robert Aitken and Sons, 1793.
Includes articles on archeology, astronomy, botany, chemistry, geology,
manufacturing, mathematics, medicine, meteorology, navigation, physics,
statistics, weights and measures, and zoology. Authors include Benjamin
Franklin, William Barton, Andrew Ellicott, Benjamin Rush, Jonathan
Williams, Benjamin Smith Barton, David Rittenhouse, Francis
Hopkinson, Caspar Wistar, Henry Muhlenberg, and Hugh Martin among
others.

801　Greenwood, Isaac (1730-1803). *Sublime Entertainment.*
[Providence: 1793]. mp: 46767 Bristol: 8353. Description of an
electrical "show" to be performed in Providence. Greenwood was a
dentist and maker of mathematical instruments; he was the son of Isaac
Greenwood (1702-1745).

802　Seaman, Valentine (1770-1817). *A Dissertation on the Mineral
Waters of Saratoga.* New York: Samuel Campbell, 1793. Evans: 26149
Austin: 1718 Hazen: 9383. Chemical analysis of spring water and a
discussion of its medicinal virtues. Seaman also includes a "topographical
description" of Saratoga and a hypothesis about how Saratoga water is
formed.

803　Woodhouse, James (1770-1809). *Observations on the
Combination of Acids, Bitters, and Astringents, Being a Refutation of the
Principles Contained in Dr. Percival's Essay on Bitters And Astringents.*
Philadelphia: Jones, Hoff & Derrick, 1793. Evans: 26496 Austin: 2087.
Description and chemical analysis of various astringents; includes several
experiments.

1794
804　Bache, William (1773-1818). *An Inaugural Experimental
Dissertation, Being an Endeavor to Ascertain the Morbid Effects of
Carbonic Acid Gas or Fixed Air on Healthy Animals, and the Manner in
Which They Are Produced.* Philadelphia: T. Dobson, 1794. Evans:
26598 Austin: 93. Medical dissertation for the University of
Pennsylvania. Bache attempts to establish the effects of "carbonic acid
gas" (carbon dioxide) on both people and animals; he includes detailed
descriptions of his experiments on animals.

805　Blakeslee, Enos. *A System of Astronomy Wherein the
Copernican System Is Refuted and the Earth Demonstrated to Be a Body
at Rest, and the Sun, Moon, and Stars to Revolve around It.* Litchfield:
Collier and Buel, [1794]. Evans: 26678. Attempt to refute heliocentric
universe; dismisses Newtonian physics as "absurd." Blakeslee's
arguments rely heavily on scripture.

806 *A Compendious System of Mineralogy and Metallurgy.* Philadelphia:
Thomas Dobson, 1794. Evans: 26801 Hazen: 2502. Extracted from
Dobson's edition of the *Encyclopedia Britannica* (see 785).

807 Drayton, John (1766-1822). *Letters Written during a Tour through the Northern and Eastern States of America.* Charleston: Harrison and Bowen, 1794. Evans: 26912 Meisel: 3: p. 358. Study of natural history and geography of the region. It contains a comparison between the schools of Boston and South Carolina which Drayton later used (as governor of South Carolina) to justify founding the University of South Carolina.

808 Mitchill, Samuel Latham (1764-1831). *Nomenclature of the New Chemistry.* New York: T. and J. Swords, 1794. Evans: 27330. Table of terms and phrases based on a work by a German chemist, Christoph Girtanner. The first column gives the new French term, the second the German equivalents from Girtanner, the third the English equivalents, and the fourth column lists synonyms and outdated terms. It ends with the announcement that Joseph Priestley is emigrating to America.

1795

809 Henfrey, Benjamin. *Sir, Having from Early Age. . . .* [Philadelphia: Snowden and McCorkle, 1795]. Evans: 28822. Syllabus for a course of public lectures on mineralogy.

810 Hosack, David (1769-1835). *Plan for Collecting the Grasses and Other Plants of the State of New York into a Herbarium.* [New York: 1795]. mp: 47459 Not included in Readex Bristol: 9144. Apparently a prospectus for founding a botanical garden; Hosack founded Elgin Botanic Garden in 1801.

811 Hosack, David (1769-1835). *A Syllabus of the Course of Lectures on Botany Delivered at Columbia College.* New York: John Childs, 1795. Evans: 28858 Meisel: 3: p. 359. Outline for a botany course; Hosack was professor of botany at the College of New York (Columbia).

812 Willis, John (fl. 1795). *An Inaugural Dissertation on the Chymical [sic] Analysis and Operations of Vegetable Astringents, with Observations on the Identity of Vegetable Acids.* Philadelphia: Alexander McKenzie, 1795. Evans: 29899. Medical dissertation for the University of Pennsylvania.

1796

813 *Analysis of Certain Parts of a Compendious View of Natural Philosophy.* Boston: William Spotswood, 1796. Evans: 29988. Textbook containing geometry, "natural philosophy," astronomy, physics, and some geology in short "epitome" form.

814 Barton, Benjamin Smith (1766-1815). *A Memoir Concerning the Fascinating Faculty Which Has Been Ascribed to the*

Rattle-Snake and Other American Serpents. Philadelphia: Henry
Sweitzer, 1796. Evans: 30037. Paper read before the American
Philosophical Society; includes a description of the "fascinating faculty"
ascribed to snakes, a review of literature and of popular belief. Barton
also describes his experiments in proving the faculty does not exist.

**815   Darwin, Erasmus (1731-1802) and Samuel Latham
Mitchill (1764-1831).** *Zoonomia or the Laws of Organic Life.*
Preface to the American edition by Samuel Latham Mitchill. New York:
T. and J. Swords, 1796. Evans: 30312. Mitchill's preface gives a brief
history of medical/biological writing. This American edition went
through more reprints than any other of Darwin's works in America.

816   Peale, Charles Willson (1741-1827). *A Scientific and
Descriptive Catalogue of Peale's Museum.* Philadelphia: Samuel H.
Smith, 1796. Evans: 30967 Hazen: 8178 Meisel: 2: p. 58.
Description of the specimens included in Peale's museum with an
introduction outlining the Linnean principles used in classifying them.
Peale's museum was the first in America devoted to natural history;
according to the DAB, "in scope and character it ranked with the notable
museums of the time."

817   Priestley, Joseph (1733-1804). *Considerations on the
Doctrine of Phlogiston and the Decomposition of Water.* Philadelphia:
Thomas Dobson, 1796. Evans: 31049. Defense of the phlogiston
theory directed to Berthollet, De la Place, Monge, Morveau, Fourcrey,
and Hallenfratz, whom Priestly describes as "the surviving answerers of
Mr. Kirwan."

818   Rush, Benjamin (1745-1813). *An Eulogium Intended to
Perpetuate the Memory of David Rittenhouse, Late President of the
American Philosophical Society.* Philadelphia: 1796. Evans: 31143.
Delivered before the American Philosophical Society, the President of
the United States, and Members of Congress. Rush gives a brief
biography of Rittenhouse, along with a description of his character and
his work, and an account of his communications with the American
Philosophical Society (of which he was president from 1791-96).

819   Saltonstall, Winthrop (1775-1802). *An Inaugural
Dissertation on the Chemical and Medical History of Septon, Ozote, or
Nitrogene and Its Combinations with the Matter of Heat and the
Principle of Acidity.* New York: T. and J. Swords, 1796. Evans:
31155. Medical dissertation for the College of New York (Columbia);
"septon" was Samuel Latham Mitchill's term for a "septic gas"; "azote"
was the common term for nitrogen.

1797
820   Maclean, John (1771-1814). *Two Lectures on Combustion,
Supplementary to a Course of Lectures on Chemistry.* Philadelphia:

Thomas Dobson, 1797. Evans: 32412. Answer to Priestley's *Considerations* (see 817); defense of the "antiphlogiston" chemistry of Lavoisier and Berthollet. Maclean was professor of chemistry and natural history at the College of New Jersey (Princeton), the first professor of chemistry in the United States in a college other than a medical school.

821 Moreau de Saint-Mery, Mederic Louis Elie (1750-1819). *A General View or Abstract of the Arts and Sciences, Adapted to the Capacity of Youth.* Translated by Michael Fortune. Philadelphia: 1797. Evans: 32505 Karpinsky: p. 115 Hazen: 7424. Textbook in question and answer format. Moreau de Saint-Mery was a French emigrant who established a printing press in Philadelphia; he was also a member of the American Philosophical Society. He later returned to France where he was an official in the Napoleonic government.

822 Priestley, Joseph (1733-1804). *Observations on the Doctrine of Phlogiston and the Decomposition of Water, Part 2.* Philadelphia: Thomas Dobson, 1797. Evans: 32719. Further defense of phlogiston chemistry (see 817).

823 Woodhouse, James (1770-1809). *The Young Chemist's Pocket Companion, Connected with a Portable Laboratory.* Philadelphia: J.H. Oswald, 1797. Evans: 33245. Description of necessary chemicals and apparatus for a beginning chemist; followed by descriptions of experiments and information about materials. The DAB calls it the "first published guide in chemical experiments for students."

1798

824 Barton, Benjamin Smith (1766-1815). *Fragments of the Natural History of Pennsylvania. Part I.* Philadelphia: Way and Goff, 1798. Evans: 35159 Hazen: 1514 Meisel: 3: p. 361. Tables giving Barton's observations of birds, plants, and other natural phenomena for several years. Greene states that the *Fragments* "[contain] much useful information about bird migrations in relation to the progress of vegetation." Barton also introduced several terms, such as "resident species" and "occasional visitant," which are still in use.

825 Barton, Benjamin Smith (1766-1815). *New Views on the Origin of the Tribes and Nations of America.* Philadelphia: John Bioren, 1798. Meisel: 3: p. 359. Evans: 33378. Discussion of the probable origins of American Indian tribes; argues for Asiatic origin based on language.

826 Smith, Thomas P. *A Sketch of the Revolutions in Chemistry.* Philadelphia: Samuel H. Smith, 1798. Evans: 34559. Speech delivered to the Chemical Society of Philadelphia; short history of the development of chemistry.

1799

827 Connecticut Academy of Arts and Sciences. *Constitution of the Connecticut Academy of Arts and Sciences.* [Hartford?: 1799]. Evans: 35344. Rules and procedures of the Academy.

828 Dewey, Sherman (1772-1813). *Account of a Hail Storm Which Fell on Part of the Towns of Lebanon, Borzah, and Franklin on the 15th of July, 1799.* Hartford: Thomas and Thomas, 1799. Evans: 35405. Description of a storm; subtitled "Perhaps never equalled by any other ever known, not even in Egypt."

829 *An Impartial Relation of the Hail-Storm on the Fifteenth of July and the Tornado on the Second of August, 1799.* Norwich: John Trumbull, 1799. Evans: 35650. Description of Connecticut hailstorm and tornado (see 828); written by a committee of residents from Bozrab, Lebanon, and Franklin, Connecticut. It includes an estimate of damages.

830 Peck, William Dandridge (1763-1822). *Natural History of the Slug Worm.* Boston: Young and Minns, 1799. Evans: 36052 Meisel: 3: p. 361. Description of life cycle and physiology of the slug worm; Peck received a gold medal and a fifty dollar premium from the Massachusetts Agricultural Society. According to the DAB the paper "described the first egg-parasite noticed in the United States."

831 Spalding, Lyman (1775-1821). *A New Nomenclature of Chemistry.* Hanover, NH: Moses Davis, 1799. Evans: 36347. Table giving old and new names of elements; divided into "orders" based on composition. Spalding was a lecturer on chemistry and materia medica at Dartmouth; the *New Nomenclature* was a translation of a French text.

832 Turner, George (fl. 1799). *Memoir on the Extraneous Fossils Denominated Mammoth Bones.* Philadelphia: Thomas Dobson, 1799. Evans: 36459. Discussion attempting to prove "mammoth bones" are "remains of more than one species of non-descript animal." Turner was a member of the American Philosophical Society.

833 [Vaughan, John (1775-1807)]. *Chemical Syllabus.* Wilmington: Bonsell and Niles, [1799]. Evans: 36609. Outline of courses, chiefly giving nomenclature.

Author Index

Subject Index

A

Agriculture 395, 398, 412, 414, 416, 456, 474, 475, 477, 485, 488, 508, 515, 518, 546, 549, 556, 557, 582, 584, 617, 618, 622, 629, 635, 636, 653, 657, 658, 675, 685, 686, 687, 689, 696, 698, 712, 713, 757, 771, 774, 799
  Bees 537, 792
  Cheese 550
  Crop Rotation 461, 536, 622, 665
  Fertilizer 670, 713
  Flax 398, 422, 449, 477
  Hemp 395, 421, 436, 449, 477, 702
  Indigo 403
  Insect Control 791, 830
  Livestock 422, 449, 477, 478, 485, 505, 584, 636, 676, 689, 712, 713
  Madder 436, 477
  Maple Sugar 519, 545
  Silk 412, 430, 432, 516, 565
  Wool 395, 422
Alcohol 122, 129
Algebra 521
Anatomy 179, 184, 352, 789
Anthrax 233
Anthropology 70, 763, 780, 788, 794, 807, 825
Apoplexy 198, 214, 343
Apothecaries 90, 137
Architecture 481, 655, 677
  See also, Engineering, Civil
Arithmetic 380, 384, 394, 459, 472, 487, 492, 495, 512, 513, 521, 526, 542, 547, 559, 560, 561, 583, 595, 609, 613, 614, 619, 645, 647, 663, 664, 668, 674, 697, 706, 710
  See also, Textbooks
Arsenic 264

Artificial Respiration 94, 104, 146, 165, 223, 231, 281, 373
Astringents, see Chemistry
Astronomy 590, 718, 725, 759, 768, 770, 771, 774, 775, 778, 784, 797, 799, 805, 813, 818
  Aurora Borealis 726
  Comets 723, 724, 737, 745, 755, 783
  Equipment 400, 734
  Meteors 726, 762
  Transits 742, 747, 751, 752
Atlas, see Geography
Aurora Borealis, see Astronomy
Autopsy 35

B

Balloons 551, 621
Bees, see Agriculture
Black, Joseph 753
Bleeding 56, 103, 266
Boone, Daniel 463
Botany 39, 50, 156, 242, 278, 310, 324, 345, 351, 545, 585, 754, 772, 789, 810, 811, 824, 828
  Pharmacopoeias, see Pharmacy
Boyle, Robert 735, 740

C

Camera Obscura 738
Canada 391, 392
Canals 470, 532, 543, 544, 610, 650, 679, 699, 708, 709
Cancer 86, 136, 178
Cantharides 361
Case Studies 19, 35, 46, 55, 81, 86, 106, 178, 200, 201, 243
Charts, Navigation, see Geography
Cheese, see Agriculture
Chemical Engineering, see Engineering, Chemical
Chemistry 156, 192, 267, 277, 327, 335, 406, 419, 442, 444, 445, 446, 526, 541, 555, 564, 605, 648, 681, 735, 753, 754,